GEOGRAPHICAL PATTERNS OF SOCIAL WELL-BEING

GEOGRAPHICAL PATTERNS OF SOCIAL WELL-BEING

(With Special Reference to Gujarat)

K. M. KULKARNI

CONCEPT PUBLISHING COMPANY, NEW DELHI-110 059

ISBN 81-7022-276-1

First Published 1990

K.M. **Kulkarni** (*b.*1949)

Published by
Ashok Kumar Mittal
Concept Publishing Company
A/15-16, Commercial Block, Mohan Garden
NEW DELHI-110059 (India)

Laser typeset by
Microtech Advance Printing Systems (Pvt.) Ltd.
H-13, Bali Nagar
NEW DELHI-110015

Printed in India by
Taj Press
3024, Gali Ahiran, B. Garh Road
DELHI-110006

DEDICATED
TO
THE PEOPLE OF GUJARAT

CONTENTS

LIST OF TABLES

LIST OF FIGURES

PREFACE

ONE of the most important tasks of social geographers in the present times is to measure spatial inequalities in social ethos. Among the various aspects of social situation, the phenomenon of social well-being is too complex to understand and measure as it includes both the objective and subjective realities and measurable and non-measurable aspects. The social well-being of any society over the world has varying degree of environmental, institutional (including cultural) and technological overtones. However, the aspect of social well-being can be better understood in terms of social needs and social attainment as they vary with geographical conditions including climatic, physiological and cultural aspects.

The present study is based on some selected indicators representing the resource making qualities of population and the socio-economic conditions of society. Despite the limitations of indicators of well-being and the inevitable generalization at the taluka level, the study contributes to the understanding of the spatial variations in the levels of social well-being in Gujarat especially so on the spatial framework of inter-taluka, inter-district and inter-regional units. The spatial patterns of composite levels of social well-being as revealed in the study may go a long way in guiding the administrators, regional and State planners, voluntary social organizations and parliamentarians in achieving the desired spatial pattern of social well-being in Gujarat.

It is my pleasure to acknowledge with a sense of gratitude the help and assistance I have received from my friends and colleagues in completing the work. Professor C.D. Deshpande, formerly Professor of Geography at the University of Bombay, remains a perennial source of inspiration to me since the days of my post-graduate and research studies. Professor (Mrs.) K. Sita and and Dr. V.S. Phadke of the University Department of Geography, Bombay; Professor Aizazuddin Ahmed of the Centre for the Study of Regional Development, Jawaharlal Nehru University, New Delhi; Professor L.S. Bhat of the Regional Survey unit, Indian Statistical Institute, New Delhi, and Professor

Ghanshyam Shah of the Centre for Social Studies, Surat, offered their comments on the preliminary work which enabled me to complete the work. I am particularly benefited by the fruitful discussions with Professor Vimal P. Shah of the Department of Sociology, Gujarat University, whose critical comments helped me in improving the methodology and quality of analysis. Geographer colleagues Professor Ambubhai Desai, Pro-Vice-Chancellor, Gujarat University, Dr. Manjula Dave of H.K. Arts College, Ahmedabad, took considerable interest in my work. I am thankful to Mr. Mudresh Purohit and Mr. Rajesh Purohit for processing the data and designing the report at their Taran Comprint, Ahmedabad. Thanks are due to Mr. Feroze Sheikh and Mr. Rajesh Joshi (Census Department) for cartographic work and to Mr. Mahendra Shah for neat typing work. I am also grateful to the University Grants Commission, New Delhi, that sponsored this minor project initially but could not be completed during its tenure 1983-85 due to certain reasons. However, I am grateful to the authorities of Gujarat University who sanctioned some research grant which enabled me to retrieve the work and present it in this form. Lastly I own the responsibility for all the weaknesses of the work if it has any.

Ahmedabad K.M. KULKARNI

1

INTRODUCTION

THE phenomena of social well-being is the most intriguing and yet challenging theme in social geography. Several scholars in the field of Geography, Sociology and Economics have studied the various dimensions of social well-being including the core-peripheral contrasts by using territorial indicators most appropriate to the regions (Drewnowski and Scott, 1968; Knox, 1975; Smith, 1977; McCracken, 1983; Ilbery, 1984). Out of these studies two important aspects emerge as sensitive and crucial as they influence significantly the pattern of social well-being. They are first, the selection of appropriate indicators reflecting the totality of social well-being and, secondly, the selection of appropriate geographical units for the purposes of a detailed study and for attaining the desired level of generalization. As regards the selection of indicators of social well-being, it is necessary to bring out the accepted socio-economic and cultural norms of different micro-regions under study. By and large, the indicators chosen should represent the major aspects such as social health, social security, social amenities and, above all, the social harmony. However, the aspect of social harmony can be recorded or measured with reasonable accuracy only by questionnaire or interview method or by devising psychometric scales. The term 'well-being' includes the various other terms such as 'level of living', 'quality of living', 'social welfare' and the level of satisfaction. However, the term 'social well-being' is a synthesis of physical well-being, material well-being, cultural and spiritual well-being though the latter aspects are either totally ignored or given inadequate importance in several studies including the present one. (due to the limitations of the project design.). The aspects of cultural and spiritual well-being are rather more difficult to measure and are yet challenging to measure and interpret.

The phenomena of social well-being include both the objective and subjective realities, measurable and non-measurable aspects of human living. There is no unanimity among researchers regarding the meaning of 'Social well-being' in terms of indicators, as the phenomena of well-being is a manifestation of regional culture. Thus, the geographical scale not only influences the selection of indicators but also the pattern of social well-being in terms of the extent of generalization. The smaller unit of study area lends to the greater efficacy of the pattern of social well-being. With the increase in the size of geographical scale the micro-differences merge with the broader and generalized pattern. Therefore, the phenomena of social well-being becomes more realistic in the micro-regional framework rather than in the meso or macro region.

Social well-being of any society over the world has cultural and technological overtones. The developed countries with their advanced technology are ahead in the level of material well-being due to the early and rapid urbanization and industrialization and concomitant change in social needs and values. But the process of industrialization and urbanization is rather slow and selective in a majority of the Third World countries resulting in a very limited change in social needs. The social values change in accordance with the social needs which are governed by the micro-geographical conditions, level of technology and cross-cultural assimilation. However, providing basic minimum needs such as food, clothing and shelter remains a priority in India and other developing countries.

Social well-being is a multifaceted phenomena which means welfare in general and health prosperity in particular. However, the phenomena of social well-being is an aggregation of an individual's level of welfare which includes the following broad aspects:

(a) *Physical well-being* : Environmental conditions such as climatic variation, local topographical variations, physical density, physical and mental health;

(b) *Material well-being* : Household assets including the modern electrical and electronic gadgets, other means of conveyance and communication;

(c) *Social well-being* : Social health, social security, social amenities and social relations including social ethos;

(d) *Economic well-being* : Employment opportunities, level of income and savings and purchasing power, consumption of goods and services;

(e) *Perception of well-being* : Attitude towards external living and non-living things, individual/social traits, experiences of real world situations and an element of subjectivity; and (*f*) *Spiritual well-being* : Religious beliefs, practices, taboos and various approaches to seek eternal truth and happiness.

The spiritual values become more important in attaining social well-being especially in societies governed by religious beliefs and practices. The traditional culture, (art literature, music, drama, dance) occupation, yoga, meditation and medicine play a significant role in social satisfaction and quality of life in a majority of the Third World countries and especially so in India. It is equally pertinent to mention here the various spiritual leaders who carried the messages of eternal truth and happiness to different parts of the affluent world. Though the cultural and spiritual aspects are important in an individual's well-being, the measurement of its share becomes difficult. However, one may attempt by choosing certain indicators of cultural and spiritual well-being.

Indicators of Social Well-being

The rapidly changing social values in modern society have direct relevance to the technological advancement at local,regional or national level in agricultural, industrial and urbanized sectors of economy. Therefore, the selection of indicators or weighing of indicators is based on the dominant activity of the region. Drewnowski, and Scott (1968) have incorporated in their study of some selected countries the seven components of levels of living: Physical, Nutrition, Shelter, Health; Cultural, Education, Leisure; Security; and Higher Surplus Income. Knox has used 53 variables for deriving the operational definition of level of living for the counties and county boroughs in England and Wales. The main constituents of his study are Housing, Health, Education, Social Status, Employment, Affluence, Leisure, Social Security, Social Stability, Demographic Structure, General Physical Environment and Democratic Participation. Smith D.M. while measuring social well-being in the United States used certain criteria such as Economic Status (income, employment, welfare); Environment (housing, streets, sewers, air pollution, open space); Health (general mortality, chronic diseases); Education (duration); Social Order (personal pathologies, family breakdown, overcrowding, public order and safety, delinquency, social belonging and democratic participation and equality. MacCracken (1983) in his study on dimensions of social well-being within seven alternative spatial frames of New Zealand has used indicators such as Male and Female income earners, Employment (male and female), Age-standardized Mortality Ratio, Level of Education and Households with telephone and television. Brain W. Ilbery (1984) has shown effectively the core-periphery contrasts in European social well-being by using (total 27 indicators) the seven major constituents, namely: Housing, Health, Education, Economic Growth, Material Well-being, Leisure and Recreation. In addition to these, several studies have been carried out including those conducted by the U.N. and the World Bank on the levels of development and social well-being in different parts of the world choosing different sets of indicators most appropriate to the region. Thus, social well-being is best reflected in the quality of living which depends, among other things, on the number, type, quality and availability of infrastructural facilities.

Keeping in view the two emerging aspects (selection of indicators and geographical scale) of several studies pertaining to the socially relevant theme like this has invoked social geographers all over the world to detect and monitor more sensitive territorial indicators for understanding the prevalent pattern of social well-being. However, the objectives of the present study are three-fold.

(1) To measure the levels of social well-being in a micro-regional framework using the regional averages (intra-regional inequalities),
(2) to examine the inter-district and inter-taluka differences in social well-being, and
(3) to examine the extent of resemblance and contrasts to patterns of urbanization.

In the present study, for reasons of data availability, the secondary and objective indicators are used. The final composite index consists of 24 variables for each of 184 talukas (of 19 districts) of Gujarat State. (For inter-district and inter-taluka study only 13 diagnostic variables are used). The major constituents of social well-being considered in the present study are:

A. *General Physical Environment*
   (1) Density of population per sq km.
   (2) Percentage of urban population.

B. *Housing*
  (3) Ratio of house to households.

C. *Employment*
  (4) Percentage of male and female workers.
  (5) Percentage of cultivators, agricultural labourers, households and manufacturing workers, others and marginal workers.
  (6) Dependency ratio.

D. *Health*
  (7) Number of beds per health unit.
  (8) Number of doctors per health unit.

E. *Education*
  (9) Number of students per primary school.
  (10) Number of students per high school.
  (11) Number of students per college.
  (12) Number of teachers per primary school.
  (13) Number of teachers per high school.
  (14) Number of teachers per college.
  (15) Teacher-student ratio at different levels.
  (16) Percentage of male and female literacy.

F. *Affluence*
  (17) Savings per household in Rs.

G. *Leisure*
  (18) Percentage of cinemagoers.
  (19) Population per cinema hall.

H. *Social Security*
  (20) Population per police station.

I. *Social Facilities*
  (21) Number of electrified villages.
  (22) Population per post office.
  (23) Population per telegraph office.
  (24) Population per post box.

Here the emphasis is mainly laid on health, education, employment, recreation, social facilities and security measures to identify the levels of social well-being with supportive field observations.

Data Base and Methodology

The present study of social well-being at the taluka level is restricted to Gujarat State and heavily relies on secondary sources of data especially the census reports of 1981 and the District Statistical Abstracts of 1980-81. The weightage is assigned to each of the variables, mentioned above, of a taluka in relation to the standards or averages of respective micro-regions by using the following functional concentration index and social well-being index (Knox's Index).

$$FCI = \frac{P_{xi} X_i}{MP\ xn}$$

where

FCI = Functional concentration index or weightage.

$P_{xi} X_i$ = Percentage or ratio of xi variable of x1 taluka.

MPxn = Mean percentage or ratio of xi variable of xn region.

Social well-being Index = $Ij = 100 \sum_{N.C}^{Rj}$ (Knox's Index).

where

Ij = Social Well-being index of district or taluka.

Rj = Sum of individual rank scores for district or taluka j.

N = Number of diagnostic variables.

C = Number of districts of talukas under study.

These methods enable us to identify the intra-regional, inter-district, inter-taluka differences in the levels of social well-being in an effective manner. The weightage thus assigned to the individual variable is added to arrive at the composite index value of social well-being. On the basis of composite index value and the population size of each of the talukas of seven micro-regions, namely (1) Kathiawar (North) and Kachchh, (2) Kathiawar Core, (3) Kathiawar Coastal, (4) North Gujarat, (5) Central Gujarat, (6) South Gujarat and (7) Eastern Hilly Region (Fig. 2.1). The four significant levels of social well-being are identified as Very High (16 talukas), High (32 talukas), Moderate (76 talukas) and low (60 talukas) (Appendix IV and V). Similarly, on the basis of Knox's Index the inter-taluka and inter-district inequalities in well-being (both demographic and socio-economic conditions) have been measured, cartographed and analysed. (For regional units and their respective administrative units refer Appendix I).

2

GEOGRAPHICAL ASPECTS AND WELL-BEING

THE micro-regional framework provides a key to the initial understanding of inequalities in the form of resource distribution and the further processes and stages involved in the levels of well-being. Drewnowski (1974) in his derivation of human well-being gives equal importance to 'resources' as to labour and technology. Therefore, the phenomenon of social well-being is better understood in the micro-regional framework which in turn reflects the variation in resource base. The seven planning regions under study are: (i) North Kathiawar and Kachchh, (ii) Kathiawar Core, (iii) Kathiawar Coastal, (iv) North Gujarat, (v) Central Gujarat, (vi) South Gujarat and (vii) Eastern Hilly Region (Fig. 2.1). These micro-regions show a marked contrast in physiographic, geological and climatic features. Thus, these aspects of environment determine the resource distribution including renewable resources: water, soil, vegetation, animal and human population; and non-renewable resources: minerals and mineral oil reserves. An attempt has been made to highlight certain geographical conditions of Gujarat which may have direct or indirect influence on social well-being.

The Location and Shape

The location and shape of Gujarat State in the Indian Union is so unique that whosoever intends to draw the sketch map of India begins to draw from the Gulf of Kachchh and the Saurashtra peninsula (Kathiawar Peninsula) for the reasons of convenience. Among the 25 States nine States have the coastal boundaries extending for nearly 6,083 kms. However, Gujarat has the longest coastline of 1600 km concealed in two Gulfs, the Gulf of Kachch and the Gulf of Cambay. The maritime location and long coastline have made considerable influence on the climate, life and activities of people.

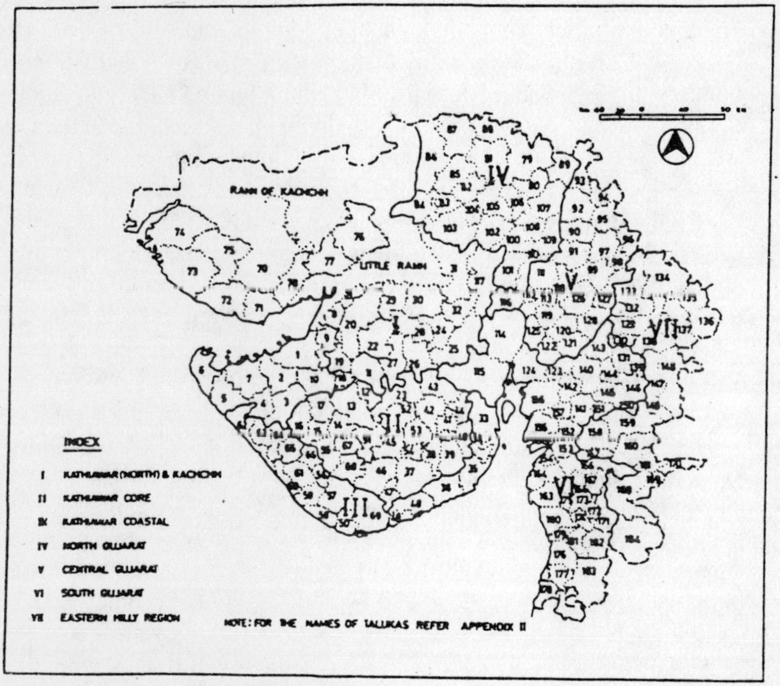

FIG. 2.1: *Micro Regions of Gujarat* (with taluka boundaries)

The geological and geomorphic features reveal the fact that the Saurashtra core and the Kachch were two separate islands and were merged into a single landmass with the mainland of Gujarat by way of the emergence of land from the sea surface. The lowlying terrain centred around Nal Sarovar reminds us of the single mass of water of the Gulf of Kachch and the Gulf of Cambay.

The entire Gujarat State is a part of the West Coast plain (one of the seven physiographic regions of India) girdled by the hill ranges in the north and east. Saputara (42 kms from Ahwa the headquarters of Dangs district) is the only well-known hill station of Gujarat being further developed for the welfare of the people. However, the people of the State get relief from the scorching summer heat by moving to Mount Abu – a hill station in the Aravalli ranges, close to the northern boundary of Gujarat but situated in Rajasthan State. Other geographical places of recreational interest include lakes and seashore beaches. Nal Sarovar of Ahmedabad district and Narayan Sarovar of Kachch district extend over a vast lowlying stretch and serve as a catchment area for the monsoon rains. The lake area provides ideal habitat conditions for the resident birds and attracts a large number of migrant birds such as White Rosy Pelicans and Flemingoes. Gujarat with its longest coastline has a few of the beautiful sandy beaches such as Ubhrat, Tithal (Valsad district) and Chorwad (Junagadh district) – the only significant sea beach resort of the entire Saurashtra coast. These resorts are remarkably pleasant even in hot summer. Pavagadh hill resort in Panchmahal district, Gir – home of Asiatic lions – and Tulsishyam hot springs in Junagadh district are the other places which offer recreational avenues to the people of Gujarat. Thus, these places provide beautiful scenic environment (coastal beach, palm groves, cool climate, bird sanctuary, wildlife sanctuary, hill station, etc.) that forms the perennial source of recreation and well-being of the people.

Climate is another most important component of physical environment that influences human health, energy and comfort. Among the various elements of climate, the temperature, sunshine and humidity exert greater influence on some of the physiological systems. However, the climatic environment constitutes all the climatic elements that together influence human comfort and well-being (Chritchfield, 1975). The food habits, clothing requirement, house design, the development of transport and communication lines and physical and mental health (including disease, hunger and morbidity) are controlled directly or indirectly by the climatic conditions. E.C.

Thom with the help of temperature-moisture index suggests the temperature limits of discomfort and uncomfort. As the index shows above 21°C, most people feel discomfort and if it exceeds 27°C then everyone feels uncomfortable. However, these indices are worked out for the temperate region and are applicable to that region. The range of temperature over time and space in Gujarat varies from 4.6°C in the month of January at Dahod to 45.2°C in the month of May at Bhuj. By and large, the various places in Saurashtra and Gujarat record high temperatures during the months of May and June. The minimum temperature ranges from 4.6°C in Dahod (Panchmahal district) to 11.1°C at Surat in the month of January. Though the temperature and humidity conditions cause discomfort and uncomfort to human living in the tropical, subtropical and equatorial regions the human adaptation and acclimatization has taken place over thousands of years. Despite this the poorest of the poor embrace death due to severe cold or heat wave. (Northern part of India and northern part of Saurashtra and Gujarat). Frequent drought and floods in Gujarat and elsewhere in India affect adversely the life and activities and thus jeopardise the well-being of people.

Resources and Activities

The salient characteristics of resource-based economic activities by a micro-regional framework has been discussed in order to understand the basis of the patterns of well-being (Resource, Labour and Technology).

The mainland of Gujarat on account of its geographical advantages and historical developments has favoured the concentration of urban population to the extent of 67 per cent. Among the seven micro-regions of Gujarat, the Eastern (hilly) region is the smallest micro-region in terms of geographical area and demographic strength. The region has less than five per cent of the total towns which were the seats of former rulers. The region has a preponderance of tribal people, particularly Bhils and Dublas, who are engaged in forest and mineral-based activities.

The central Gujarat plain with Ahmedabad urban agglomeration as its core accommodates 31 per cent of the urban population of Gujarat. The region is agriculturally very productive, industrially most developed and displays an excessive concentration of capital forming enterprises. The Charotar area which is a part of the central Gujarat plain is agriculturally most prosperous in terms of intensity of cropping, productivity and income generation. The industrial activities and the associated land uses are quite prominent in the Central Gujarat plain which includes among the other locations Ahmedabad-Vadodara industrial belt. The industrial region is characterized by a high concentration of urban population, intensive farming and high rural density of population and diversified industrial base (textiles, chemical, engineering and dairy industries). Amul Dairy at Anand in Kheda district has attained prominence at the national and international levels on account of its success in cooperative milk production. The industrial development around Vadodara has acquired importance at the national level on account of its petro-chemical complexes. The large number of pharmaceutical and chemical industries have come into existence in the recent past, which has slowly changed the functional character of this princely town.

The South Gujarat plain shows a higher proportion of workers engaged in the manufacturing, trading and commercial activities. The region is endowed with rich reserves of petroleum and natural gas; fertile black soil, abundant rainfall and surface water and moderately dense forest cover. Over and above the resource base, the region is situated midway between the industrial belts of Ahmedabad in the north and the Bombay in the south and therefore enjoys the locational advantages. The industrial development of the area is associated with the oil-fields of Ankleshwar, the thermal power house at Utran and many other small-scale industries in and around Surat. The forest wealth of Dangs and Valsad has influenced Bilimora to become an important centre of forest-based industries. A significant development in the crop pattern of the area in the recent past includes the marked concentration of plantation crops such as banana and chickoo.

In terms of industrial activities, the northern part of North Gujarat plain is the least developed part in Gujarat. The southern part of the North Gujarat region has agro-based industries such as ginning and pressing, cotton-seed oil and cotton textiles (Kadi, Kalol, Mehsana). Three-fourths of the central part of North Gujarat is under cultivation. Some of the towns are well developed as market centres for agricultural commodities such as cummin, isabgul and dilm. Sidhpur and Unjha towns are the national markets for cummin and isabgul. The region is also endowed with large deposits of china clay which has favoured the development of ceramic industries.

The whole of the Saurashtra region accounts for 33 per cent of the urban population of Gujarat. The region has rich reserves of industrially useful minerals like china clay, limestone, gypsum, lignite, bauxite, calcite, salt and brine. These raw materials are used in various industries such as cement (Dwarka, Sikka). Chemicals (Mithapur, Kharaghoda, Porbandar and Dhrangadhra). The agricultural productivity is low on account of saline soil and brakish water especially in the northern peripheral region of Saurashtra. The primary activity is more significant in the Kathiawar-core region which has the highest percentage of workers in primary activities. Groundnut is the dominant crop of the region and agriculturally it is more productive here than in the Rajkot plateau. There are also well-known centres of groundnut oil industry in this region. The region has some of the well-developed princely towns.

The Kathiawar coastal region is characterized by a high amount of rainfall, high agricultural productivity and dense population. The calcareous deposits near the coast particularly milliolite have promoted the cement industry at Porbandar. Besides cement, there are quite a few industrial units manufacturing various chemicals and processed food.

Thus, it could be seen from the preceding discussion that the Central Gujarat region is marked by intensive industrial, urban and commercial activities; the South Gujarat by industrial (forest, petroleum-based) activities; and the North Gujarat by agricultural and commercial activities. The economic activities in the Saurashtra region include mainly agricultural, mining, fishing and industries related to these primary activities.

3

SOME OF THE RESOURCE-MAKING CHARACTERISTICS OF POPULATION

THE phenomena of social well-being, though subjective, can be measured by studying several indirect indicators. Among them the demographic structure or human resource, socio-cultural and economic indicators are important. The resource-making characteristics of human population include density of population, level of literacy, proportion of working population, proportion of females and dependents, proportion of urban population and proportion of houses to households. These aspects of population reflect to a greater extent the territorial inequalities in the levels of well-being (Table 3.1 and Table 3.2).

Density of Population

The density of population is one of the important indicators of geographical synthesis and economic activities. The density of population is a better surrogate for carrying capacity of the land in rural areas. Thus, indirectly it brings out the productivity of land. In urban areas, the high density of population reflects, among other things, the accessibility to various social amenities (school, office, market, hospital, post and telegraph, bus and railway station, parks, playgrounds etc.) contributing to better level of well-being. The intra-urban high density core areas have shown a better level of well-being in Ahmedabad city than the peripheral areas (Kulkarni, 1984).

Among the seven micro-regions of Gujarat the Central Gujarat region shows very high density of population (441 persons per sq km as compared to 221 for India). The spatially contiguous region of South Gujarat shows the density of 320 persons per sq km due to the impetus provided by both the metropolitan region of Bombay and the Ahmedabad-Vadodara industrial belt. The remaining micro-regions have the density of population less than 221 persons per sq km which is the national average (Table 3.3). The average density of population in the State (174 persons per sq km) too is much less than the national average on account of the large area under arid and semi-arid conditions covering Saurashtra peninsula, Kachchh and North Gujarat.

TABLE 3.1: Region-wise Demographic Structure — 1981

Micro-regions	*Total Geographical area (sq km)*	*Total population*	*Total urban area (sq km)*	*Total urban population*	*Total number of towns*	*Percentage of urban population of Gujarat*
I. Kathiawar (north and Kachchh	47442.1	4848516	1348.2	1714999	45	16.26
II. Kathiawar Core	18301.8	3131970	649.5	838104	30	7.97
III. Kathiawar Coastal	14257.9	2920124	690.2	948502	27	9.05
IV. North Gujarat	21426.8	4216701	357.4	655588	19	6.27
V. Central Gujarat	17139.9	7562913	705.8	3380525	49	31.45
VI. South Gujarat	24376.2	7800431	906.3	2744393	75	26.34
VII. Eastern Hilly Region	18381.6	3605144	107.2	319542	10	2.66
Gujarat State	161326.3	34085799	4764.6	10601653	255	100.00

Note: Total geographical area figures supplied by the Surveyor General do not tally with the total area of the State as the latter is compiled on the basis of tahsil area. In certain cases (Kachchh, Jamnagar, Bhavnagar districts) the entire geographical area is not divided among its villages.

Among the 19 administrative districts of the State Ahmedabad district and the contiguous one taluka district of Gandhinagar show a high density of 445 persons per sq km. The other contiguous districts having the density of population more than the national average are Mehsana-28:

(North Gujarat), Kheda-419, Panchmahal-262 (Central Gujarat) and Vadodara-328, Surat-326, Valsad-338 (South Gujarat). The low density of population is observed in Kachch district (23 per sq km) and Dangs (64 per sq km) due to inhospitable geographical conditions. Among the 184 talukas of the State, Ahmedabad City taluka records the highest density of population (7,966 persons per sq km) and the lowest density is observed in Lakhpat taluka and Kachch district (15 per sq km). Thus, the inter-district and intra-district variation in the density of population is quite significant.

TABLE 3.2: District-wise Demographic Structure — 1981

	District	*Total Geographical area (sq km)*	*Total population*	*Total urban area (sq km)*	*Total urban population*	*Number of towns*
1.	Jamnagar	14125	1393076	427.3	521592	17
2.	Rajkot	11203	2093094	253.3	864322	12
3.	Surendranagar	10489	1034185	386.4	296987	11
4.	Bhavnagar	11155	1879340	402.0	625574	16
5.	Amreli	6760	1079097	281.3	220342	12
6.	Junagadh	10607	2100709	535.7	639801	22
7.	Kachchh	45652	1050161	332.8	274372	10
8.	Banaskantha	12703	1667914	90.4	144168	5
9.	Sabarkantha	7390	1502284	121.8	148710	8
10.	Mehsana	9027	2548787	267.0	511420	14
11.	Gandhinagar	649	289088	56.8	62443	1
12.	Ahmedabad	8707	3875794	493.0	2781078	30
13.	Kheda	7194	3015027	328.4	606355	18
14.	Panchmahals	8866	2321689	113.1	257552	9
15.	Vadodara	7794	2558092	242.1	950527	19
16.	Bharuch	9038	1296451	106.8	241509	8
17.	Surat	7657	2493211	182.6	1066039	21
18.	Valsad	5244	1774136	143.8	388862	22
19.	Dangs	1764	113664	—	—	—
	Gujarat State	196024	34085799	4764.6	10601653	255

Slow Social Change

The degree of urbanization is positively associated with the quality of living. The urban areas, by and large, are better served with social amenities compared to the rural areas. Thus, marked rural-urban contrasts in terms of the quality of environment, socio-economic inputs

TABLE 3.3: Some Demographic Indicators — 1981 (regionwise)

Micro-region	*Density of population (per sq km)*	*Rank*	*Urban density (per sq km)*	*Rank*	*Percentage of urban population*	*Rank*	*Percentage female literacy*	*Rank*	*Percentage of work-force*	*Rank*	*Number of households per 1000 houses*	*Social well-being index*
I Kathiawar (North) and Kachchh	102	7	1272	7	35	2.5	27	4	32	4	1007	70.00
II Kathiawar Core	171	6	1290	6	27	5	33	2	32	3	1011	62.86
III Kathiawar Coastal	205	3	1374	5	32	4	25	5	31	5	1006	62.86
IV North Gujarat	197	4	1834	4	16	6	22	6	30	7	1010	77.14
V Central Gujarat	441	1	4790	1	45	1	29	3	30	6	1013	34.29
VI South Gujarat	320	2	3028	2	35	2.5	35	1	38	1	1036	24.29
VII Eastern Hilly Region	196	5	2981	3	9	7	17	7	37	2	1009	68.57
Gujarat State	174		2225		31		32		32		1010	

Note: Ranks are given on the basis of the figures upto two decimals.

TABLE 3.4: Some Demographic Indicators — 1981 (districtwise)

	Districts	*Density of population (per sq km)*	*Rank*	*Urban density (per sq km)*	*Rank*	*Percentage of urban population*	*Rank*	*Percentage female literacy*	*Rank*	*Percentage of work-force*	*Rank*	*Average number of persons employed in working factories*	*Social well-being index*
	1	*2*	*3*	*4*	*5*	*6*	*7*	*8*	*9*	*10*	*11*	*12*	*13*
1.	Jamnagar	99	16	1221	13	37	4	30	11	31	12	18854	58.95
2.	Rajkot	187	11	3412	4	41	3	39	2	31	10	37791	31.58
3.	Surendranagar	99	17	769	18	29	8	26	15	33	7	18848	68.42
4.	Bhavnagar	168	12	1556	11	32	7	28	13	32	8	17369	53.68
5.	Amreli	160	13	783	17	20	12	32	10	31	11	3766	66.32
6.	Junagadh	198	10	1194	14	31	6	30	12	30	17	22897	62.11
7.	Kachchh	23	19	824	16	26	9	27	14	32	9	7709	70.53
8.	Banaskantha	131	15	1595	10	9	18	11	19	30	15	1147	81.05
9.	Sabarkantha	203	9	1221	12	10	17	26	16	30	14	8523	71.58
10.	Mehsana	282	7	1915	8	20	16	35	7	29	19	19669	60.00
11.	Gandhinagar	445	1	1099	15	22	11	39	3	29	18	3437	50.53
12.	Ahmedabad	445	2	5641	2	72	1	46	1	30	16	253072	23.16
13.	Kheda	419	3	1846	9	20	13	34	8	31	13	39843	48.42
14.	Panchmahals	262	8	2277	6	11	15	15	18	33	6	8131	55.79
15.	Vadodara	328	5	3926	3	37	5	37	6	33	5	88900	25.26
16.	Bharuch	143	14	2261	7	19	14	33	9	37	3	15620	49.47
17.	Surat	326	6	5838	1	43	2	38	5	40	2	58668	16.84
18.	Valsad	338	4	2704	5	22	10	38	4	36	4	43648	28.42
19.	Dangs	64	18	—	—	—	—	21	17	40	1	141	63.16
	Gujarat State	174		2225		31		32		32		668033	

Note: Ranks are given on the basis of two decimal figures.

and the levels of living influence the pattern of well-being. Social well-being of any society over the world has cultural and technological overtones. The advanced societies with their high level of technology are ahead in terms of material well-being. The early and rapid urbanization and industrialization have brought in changes in social values. But the process of social change is rather slow and selective in a majority of the Third World countries on account of limited urbanization and industrialization.

Nearly 58 per cent of urban population of Gujarat is concentrated over (124 towns) the Central and South Gujarat regions (Table 3.1). The Kathiawar (north) and Kachchh region has 45 towns which provide shelter to about 16 per cent of the urban population. The other four regions have less than 10 per cent of urban population in each region. The inter-district variation in the proportion of urban population ranges from 8.6 per cent in Banaskantha district to 71.8 per cent in Ahmedabad district. However the Dangs district does not have any urban population (Table 3.4). The inter-taluka variation is more pronounced with Chikhli taluka (Valsad district) recording 2.8 per cent and Ahmedabad City taluka 99.1 per cent. The other highly urbanized talukas of the state are Chorasi – 85.5 per cent (Surat city), Vadodara – 79.6, Rajkot – 78.6, Bhavnagar – 77.8 and Jamnagar – 70.7. (Refer Appendix II).

Nearly 6.8 per cent of the total urban area of the State accommodates 36.5 per cent of the total urban population of the State covering only nine talukas out of a total of 184. These talukas have the density of urban population ranging from 61,400 persons per sq km in Padra taluka (near Vadodara) to 8800 persons per sq km in Dahod taluka. The next density category of 5000-7500 persons per sq km covers nearly 14 talukas with 7.6 per cent of the total urban area and 20.8 per cent of the total urban population of the State. In this category the urban density varies from 5,014 persons per sq km in Dhoraji to 7,490 persons per sq km in Jamnagar. Thus, the concentration of urban population is in the Central and South Gujarat region and to a limited extent in the Kathiawar (north) and Kachchh region.

Literacy an Important Indicator

The proportion of literate population is considered to be one of the important territorial indicators for measuring the level of community living or social well-being. It is believed that literacy in general brings about much desired public awareness especially by way of their effective participation in the developmental activities. However, the percentage of literacy alone as an indicator of social progress and well-being need not always show an encouraging pattern of well-being. The other factors such as level of education, diversification and specialization of education significantly contribute to the well-being of society. On account of educational facilities, communication media network, cosmopolitan social structure and selective but effective interaction in urban areas, the proportion of the literates and the educated may be more in urban areas.

The micro-regional variation in the proportion of literate male population is less accentuated (between 34 per cent and 65.5 per cent) as compared to the variation in the literate female population. (11.4% and 45.5%). The Central Gujarat region has 57 per cent literate population and the contiguous area of South Gujarat has 56 per cent literate people. The female literacy is supposed to be a better indicator of family well-being as it influences the quality of living. The female literacy is relatively high in the South Gujarat region (35 per cent of female population). The Central Gujarat region though highly urbanized and industrialized records only 29 per cent female literacy which is less than the female literacy of the Kathiawar core region (33 %). The remaining regions show female literacy of about 25 per cent and below. The Eastern Hilly and tribal region has the lowest proportion of literate females (16 %). Among the administrative units Ahmedabad district has the highest proportion of male (65.5%) and female literates (45.5%) and Banaskantha the lowest (34.1 per cent male literates and 11.4 per cent female literates). The talukas recording maximum and minimum proportion of male literate population in Gujarat are in Vadodara district. Vadodara taluka has the highest proportion of male literate population (72%) and Chota Udaipur taluka of the same district has the lowest (21.9%). Ahmedabad City taluka has 54.9 per cent female literates and Vav taluka of Banaskantha district has a low of 4.3 per cent. The Z Scores* computed for the literacy

---

$$Z = \frac{x-\bar{x}}{S}$$

Ahmedabad +1.30, Gandhinagar +0.82, Kheda +0.56, Rajkot +0.53 Vadodara +0.49, Valsad +0.34, Surat +0.32, Mehsana +0.32.

Bharuch –0.09, Amreli –0.17, Jamnagar –0.33, Junagadh –0.36, Bhavnagar –0.41, Sabarkantha –0.41, Surendranagar –0.65, Kachchh –0.87, Dangs –1.46, Panchmahals –1.64 and Banaskantha –2.17.

level show positive values for Ahmedabad, Gandhinagar, Kheda, Rajkot, Vadodara, Valsad, Mehsana and Surat districts and negative values for Bharuch, Amreli, Jamnagar, Junagadh, Bhavnagar, Sabarkantha, Surendranagar, Kachchh, Dangs, Panchmahals and Banaskantha in a decreasing order. Thus, high literacy level prevails in the Central and South Gujarat regions and also in Rajkot and Mehsana districts.

Labourforce in Formal Sectors

Human resource is the basic requirement for attaining desired economic objectives. The proportion of labourforce in different sectors of economy determine the level of economic development as well as material well-being. By and large, the proportion of workforce in all the micro-regions remains around 30 to 35 per cent of the population. The percentage of male workers varies between 50 and 55 whereas the proportion of female workers varies between 7 (North Gujarat) and 18 (South Gujarat). The South Gujarat region has a relatively higher proportion of male and female workers. The district employment pattern in favour of males emerges as a result of social discrimination by sexes. The inter-district variation in the proportion of workforce is between 29.1 per cent (Mehsana district) and 39.8 per cent (Dangs). The proportion of male workforce varies from 49.2 per cent in Mehsana and Gandhinagar districts to 57.7 per cent in Surat district and the female workforce varies from 6.8 per cent in Banaskantha to 22.8 per cent in Surat district. Among the talukas of Gujarat, Dediapada (Bharuch district) has the highest proportion of total (48.0 %) workers and total female workers (40.1%) and Uchchhal taluka of Surat district the highest male workforce (60.5%). Similarly the lowest proportion of total workforce and female workforce is observed in V. Mahal of Sabarkantha district and the lowest male workforce in Bhavnagar taluka. Thus, it is observed from the preceding discussion that the concentration of workforce is in South and Central Gujarat, parts of North Gujarat and the Eastern Hilly region (Female workforce).

In all the micro-regions the proportion of workers engaged in agricultural activity (both cultivators and agricultural labourers) is more than 63 per cent of their working population. Agriculture is a dominant activity in the Eastern hilly region as there are nearly 84 per cent workers engaged in it. The South Gujarat region is the only region having a higher proportion of agricultural labourers compared to the cultivators. This is mainly because of the intensive agriculture practiced in the usually small-sized holdings of the region. The average number of workers employed in working factories is more in Ahmedabad, Vadodara, Surat, Valsad and Kheda districts (Table 3.4).

The proportion of dependants is relatively more in the North Gujarat, Kathiawar Coastal and Central Gujarat regions. The dependency ratio is less than two in the Eastern Hilly region (1.44). The dependent members are the burden to the family exchequer and divide the total well-being.

Housing A Basic Need -- Unfulfilled

The dwelling unit is one of the three basic needs of human beings which remains unfulfilled. Housing facility is one of the vital aspects of civilized life. It is not only an indicator of social welfare but also of cultural and economic well-being. However, census statistics on housing in India does not differentiate between the houses used for residential purpose and non-residential purpose. Therefore, relating the total number of houses to the number of households may not present any true situation of the housing stock. Keeping in view the limitation of the ratio of houses to households, the spatial pattern is analysed. Among the micro-regions, South Gujarat region shows greater shortage of houses (1,000 houses to 1,036 households). Other regions show an excess of 7 to 13 households per 1000 houses (Table 3.3). The inter-district variation in the shortage of houses ranges from 5 (Jamnagar, Surendranagar, Amreli) to 38 (Surat) houses per 1,000 households. The magnitude of the shortage of houses per 1,000 households is the highest in Palsana taluka (195 houses per 1,000 households) of Surat district and the minimum in Kankrej (Banaskantha district) and Dhandhuka (Ahmedabad district) talukas (3 per 1,000 households). Thus, it is observed that the shortage of houses in relation to the total number of households is maximum in the South Gujarat region that too in Surat district and particularly in Palsana taluka. This is perhaps due to the increasing industrial activities and concomitant exodus of migrants into the region.

The spatial pattern of index values displays a marked inequality with multi-core periphery relationship. The micro-regional analysis shows two core areas with better social well-being (low index value). They are South Gujarat (24.29 value) and the Central Gujarat region (34.29 value). The remaining regions forming the peripheral area show

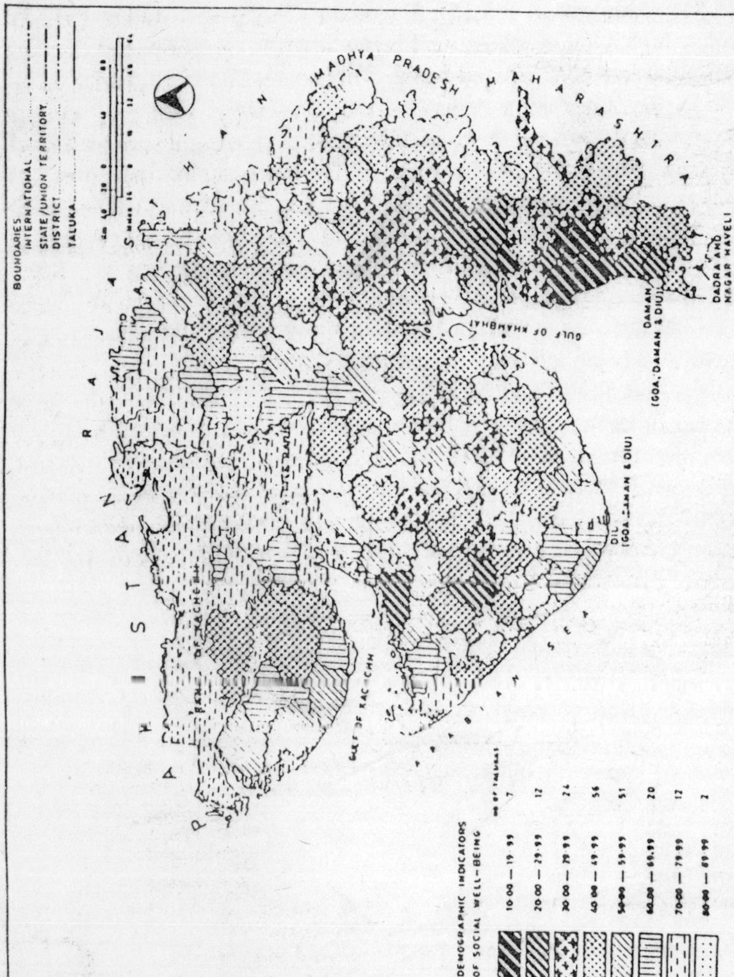

FIG. 3.1: *Patterns of Social Well-being in Gujarat*
(based on demographic index value)

a marked low level of social well-being (Table 3.3). The inter-district pattern of index values emerges with four core areas, namely, Surat (16.84), Ahmedabad (23.16), Vadodara (25.26) and Rajkot (31.58), with a high level of social well-being whereas the peripheral districts indicate a low level of well-being (Table 3.4). The intra-district variation in the distribution pattern of index values is shown by talukas (Appendix V). Among the total of 19 talukas showing a very high level of social well-being on the basis of the demographic indicators, 13 talukas are from Surat (6), Valsad (4) and Bharuch (3) districts of South Gujarat. Another three talukas of Vadodara district of Central Gujarat and one taluka each of Rajkot and Jamnagar of the North Kathiawar region show a very high level of social well-being (Table 3.5). The majority of the talukas showing very high and high levels of social well-being are from the core area of the South Gujarat, Central Gujarat and North Kathiawar regions (Fig. 3.1). Similarly, the large number of talukas of low and very low levels of social well-being are from the peripheral districts of Banaskantha, Sabarkantha, Rajkot, Jamnagar, Junagadh, Amreli and Panchmahals. These areas are marked by adverse geographical conditions such as semi-arid and arid climate, sandy-loamy soil and rocky hilly terrain in certain parts. (Low productivity of the soil.)

TABLE 3.5: Distribution of Talukas by Index Values

Social well-being index	*Number of Talukas*		*Level of well-being*
Less than 30.00	19	(Surat-6, Valsad-4, Bharuch-3 Vadodara-3, Rajkot-2, Jamnagar-1)	Very high
30.01 - 40.00	24		High
40.01 - 50.00	56		High-moderate
50.01 - 60.00	51		Low-moderate
60.01 - 70.00	20		Low
More than 70.00	14	(Banaskantha-6, Sabarkantha-3, one each from Kuchchh, Rajkot, Jamnagar, Panchmahals, and Mehsana)	Very low

Thus, the present section brings out the pattern of social well-being as seen through the demographic indicators. It is a partial viewpoint of the overall level of social well-being. However, the pattern can be compared and combined with the index values derived on the basis of certain socio-economic and cultural indicators discussed in the following section.

4

ASPECTS OF SOCIO-ECONOMIC CONDITIONS AND LEVELS OF LIVING

BASICALLY, social well-being has a cumulative effect on an individual's aspirations, needs, achievements and feeling of satisfaction. Quite often these aspects are well reflected in the visual and factual circumstances of an individual or a group of individuals. Though there are certain reservations in using the official statistics (as they do not fully represent the true situations) it becomes inevitable to depend on such a kind of data due to lack of suitable alternatives. The official statistics on number and capacity of various public distribution or service systems such as education, health care, police protection, rural electrification, transport and communication network and other entertainment services broadly indicate the territorial pattern of the level of living, quality of life and ultimately the level of social well-being. The technologically advanced countries with the culture of material affluence are a way ahead in the level of social well-being. In contrast, the developing countries of the Third World, including India, show a varying degree of low level of social well-being. However, the collective level of satisfaction diminishes with the higher order need or source. In other words, the majority of people in the Third World derive greater share of their satisfaction by necessary goods and services and only a small proportion of population derives satisfaction from all the sources such as necessities, comforts and luxuries of life (Drewnowski and Scott, 1968).

Keeping in view the minimum needs of an individual and the society, the territorial social indicators are chosen in the present study to ascertain the spatial patterns of social well-being by micro-regions, inter-district and intra-district units. Some of the social indicators chosen for a detailed analysis include the following:

(1) Number of primary schools per 10,000 primary school students.
(2) Number of primary teachers per 10,000 primary school students.

(3) Number of doctors per 100,000 population.
(4) Number of hospital beds per 10,000 population.
(5) Number of post-boxes per 10,000 population.
(6) Percentage of towns and villages with electric power.
(7) Length of road network per 1,000 sq km area (in kms)
(8) Number of cinema halls per 100,000 population.

Primary Schools

The primary school is considered as an ubiquitous facility. The local self-government bodies and the primary school boards are committed to enforce free and compulsory primary education for all eligible children. Thus, the programme has benefited rural and tribal children in acquiring the basic training of reading, writing and speaking in regional language. Whether the facility is need-based or programme-based it certainly makes a positive impact on the social environment. The process of learning among children begins in a formal set-up at the primary school though it does not insist for any specific condition or set-up. The acquired knowledge and gained experience by way of formal and informal processes of learning shape the destiny of an individual and his way of living. Therefore, the presence of primary schools and their number in relation to the number of enrolled primary students suggests the spatial disparities or inequalities in the provision of the facility to the school-going children (Tables 4.1 and 4.2). Among the seven micro-regions of the State the eastern hilly region shows the highest number of primary schools (92) per 10,000 primary students and the lowest of 35 primary schools per 10,000 students can be observed in the Kathiawar core region (Table 4.3). The inter-district variation also shows the higher number of primary schools in the peripheral, tribal and backward districts, namely, Dangs-141, Panchmahals-83, Sabarkantha-76, Banaskantha-73 and Kachchh-77 (Table 4.4). This kind of spatial pattern is perhaps largely attributed to the overzeal of the administrator-planners in providing the facility of primary school in these peripheral districts. It is also true that because of the nature of terrain, the special privileges given to the tribal society, and the distribution of villages, the region has more primary schools. The hilly terrain of the east and north and the desert conditions of the north-west (Kachchh region) have favoured the sparse distribution of villages and poor inter-connectivity among them. This peculiar condition has favoured the individual villages to have an independent

TABLE 4.1: Micro-regional Pattern of Socio-economic Variables

Micro-regions	Number of primary schools	Number of primary students	Number of primary teachers	Number of doctors	Number of hospital beds	Number of post boxes	Number of settlements electrified	Total length of roads (in kms)
I. Kathiawar (North) and Kachchh	3613	648395	16826	457	3815	3547	2074	10444
II. Kathiawar Core	1858	529930	13171	218	2054	2302	1355	5098
III. Kathiawar Coastal	1735	420519	9698	168	1324	1908	1096	3332
IV. North Gujarat	3102	587257	13720	247	1449	3025	1893	5503
V. Central Gujarat	4876	982201	29000	827	5617	4541	2215	10293
VI. South Gujarat	5639	1167061	30806	526	4531	6832	2674	12230
VII. Eastern Hilly Region	4111	440743	12504	176	1098	4054	1223	7989
Gujarat State	24934	4776106	125725	2619	19888	26209	12530	54889

Note: Compiled from District Statistical Abstracts — 1980-81.

TABLE 4.2: Inter-district Pattern of Socio-economic Variables

Districts	*Number of primary schools*	*Number of primary students*	*Number of primary teachers*	*Number of doctors*	*Number of hospital beds*	*Number of post bozes*	*Number of settlements electrified*	*Total length of roads (in kms)*
1. Jamnagar	996	188925	5472	128	1397	951	554	1908
2. Rajkot	1257	344563	8663	166	1334	1155	711	2764
3. Surendranagar	715	129152	3231	109	875	877	451	2736
4. Bhavnagar	1143	252854	6177	116	1784	1324	760	2781
5. Amreli	728	187095	4286	79	422	853	464	2128
6. Junagadh	1202	335083	7687	135	894	1423	876	2080
7. Kachchh	951	118538	3057	103	431	995	585	3894
8. Banaskantha	1574	216788	4895	98	336	1501	867	3824
9. Sabarkantha	1642	214810	6984	118	893	1924	1134	3686
10. Mehsana	1528	370469	8825	149	1113	1524	1026	1679
11. Gandhinagar	128	40915	1076	43	175	214	52	295
12. Ahmedabad	1584	489226	12058	631	3743	1298	588	4830
13. Kheda	2257	404531	14099	75	1218	1701	811	4734
14. Panchmahal	2231	270175	7611	94	544	2088	703	4670
15. Vadodara	1955	359564	9112	193	1935	1923	1061	2573
16. Bharuch	1327	192741	5335	67	488	1607	550	2845
17. Surat	1915	365501	8720	192	1432	3033	716	3700
18. Valsad	1443	269828	7622	110	725	1451	588	3537
19. Dangs	358	25348	815	13	149	357	33	1025
Gujarat State	24934	4776106	125725	2619	19888	26209	12530	54889

Note: Compiled from District Statistical Abstracts, 1980-81.

TABLE 4.3: Regionwise some Diagnostic Indicators of Well-being

Micro-regions	*No. of primary schools per 10,000 students*	*No. of primary teachers per 10,000 students*	*No. of doctors per 100,000 population*	*No. of beds per 10,000 population*	*No. of post-boxes per 10,000 population*	*Percentage of towns and villages electrified*	*Length of road per 1,000 sq km area*	*No. of cinema halls per 100,000 population*
I. Kathiawar North	60	250	9	8	7	74	220	1.59
and Kachchh	(2)	(4)	(2)	(1)	(5)	(6)	(7)	(3)
II. Kathiawar	35	238	7	7	5	86	279	1.60
Core	(7)	(5)	(4)	(3)	(7)	(1)	(4)	(2)
III. Kathiawar	41	227	6	5	8	78	234	1.47
Coastal	(6)	(6)	(6)	(5)	(3)	(4)	(6)	(4)
IV. North	56	222	6	3	7	78	257	1.39
Gujarat	(3)	(7)	(5)	(6)	(4)	(3)	(5)	(5)
V. Central	54	303	11	7	6	79	578	1.18
Gujarat	(4)	(1)	(1)	(2)	(6)	(2)	(1)	(6)
VI. South	54	263	8	6	8	77	502	1.60
Gujarat	(5)	(3)	(3)	(4)	(2)	(5)	(2)	(1)
VII. Eastern	92	270	5	3	11	36	435	0.95
Hilly Region	(1)	(2)	(7)	(7)	(1)	(7)	(3)	(7)

Note: Ranks are given on the basis of two decimal figures.

TABLE 4.4: District-wise some Diagnostic Indicators of Well-being

Districts	*No. of primary schools per 10,000 students*	*No. of primary teachers per 10,000 students*	*No. of doctors per 100,000 population*	*No. of beds per 10,000 population*	*No. of post-boxes per 10,000 population*	*Percentage of towns and villages electrified*	*Length of road per 1,000 sq km (in km)*	*No. of cinema halls per 100,000 population*
1	*2*	*3*	*4*	*5*	*6*	*7*	*8*	*9*
1. Jamnagar	53 (11)	290 (4)	9 (6)	10 (2)	7 (14)	78 (7)	135 (18)	1.36 (13)
2. Rajkot	36 (16)	251 (10)	8 (7)	6 (8)	6 (17)	81 (5)	247 (15)	1.86 (2)
3. Surendranagar	55 (8)	250 (11)	11 (4)	8 (5)	8 (7)	68 (12)	261 (13)	1.55 (8)
4. Bhavnagar	45 (13)	244 (13)	6 (15)	9 (4)	7 (13)	85 (4)	249 (14)	1.60 (6)
5. Amreli	39 (15)	229 (17)	7 (10)	4 (16)	8 (10)	76 (9)	315 (10)	1.85 (3)
6. Junagadh	36 (17)	229 (16)	6 (13)	4 (12)	6 (18)	77 (8)	196 (16)	1.71 (5)
7. Kachchh	77 (3)	258 (9)	10 (5)	4 (15)	10 (5)	64 (13)	85 (19)	1.52 (10)
8. Banaskantha	73 (5)	226 (18)	7 (12)	2 (19)	9 (7)	64 (14)	301 (12)	0.90 (15)
9. Sabarkantha	76 (4)	325 (2)	8 (8)	6 (9)	13 (2)	80 (6)	500	0.73 (16)

(*Contd.*)

TABLE 4.4: *(Contd.)*

1	*2*	*3*	*4*	*5*	*6*	*7*	*8*	*9*
10. Mehsana	41 (14)	238 (15)	6 (16)	4 (11)	6 (15)	94 (2)	186 (17)	1.22 (14)
11. Gandhinagar	31 (19)	263 (7)	15 (2)	6 (8)	7 (12)	69 (11)	455 (7)	0.69 (17)
12. Ahmedabad	32 (18)	246 (12)	16 (1)	10 (3)	3 (19)	86 (3)	346 (8)	1.78 (4)
13. Kheda	56 (7)	348 (1)	2 (19)	4 (14)	6 (16)	100 (1)	658 (2)	1.43 (12)
14. Panchmahal	83 (2)	208 (19)	4 (18)	2 (18)	9 (6)	37 (18)	527 (4)	0.60 (18)
15. Vadodara	54 (9)	259 (8)	7 (11)	7 (6)	8 (11)	62 (15)	330 (9)	1.45 (11)
16. Bharuch	67 (6)	266 (6)	5 (17)	4 (17)	12 (3)	49 (17)	315 (11)	1.54 (9)
17. Surat	52 (12)	239 (14)	8 (9)	6 (10)	12 (4)	58 (16)	483 (6)	2.45 (1)
18. Valsad	53 (10)	282 (5)	6 (14)	4 (13)	8 (9)	70 (10)	674 (1)	1.58 (7)
19. Dangs	141 (1)	322 (3)	11 (3)	13 (1)	31 (1)	11 (19)	581 (3)	— (—)

primary school of their own. Among all the talukas of Gujarat, Lakhpat taluka of Kachchh district shows a high number of primary schools (145) per 10,000 primary school-going children, and the minimum number of schools (13) is observed in Upleta taluka of Rajkot district. However, the average number of students per primary school is maximum (749 students) in Upleta taluka and minimum (73) in Lakhpat taluka. In terms of the absolute number of primary schools Panchmahal district has the highest number (2,307) primary schools) closely followed by Kheda district with 2,175 primary schools. Gandhinagar and the Dangs being smaller districts have 136 and 360 primary schools respectively. The whole of Gujarat State had 25,074 primary schools during 1980-81. Thus, the Gujarat plain region showed a higher number of primary schools along with their higher average number of students. In the Saurashtra peninsula, Rajkot, Junagadh and Bhavnagar districts showed a relatively better situation in the case of primary schools.

Teacher-Student Ratio An Important Indicator

The primary school teachers play a significant role in inculcating lasting social values in the minds of children. Therefore, it is necessary to give adequate importance to their number, training, benefits and well-being. They in turn promote and generate well-being of society by way of imparting knowledge. There are about 1,29,916 teachers in primary schools in Gujarat State who teach nearly 51,05,421 students. Nearly 21.42 per cent of the total primary teachers and 21.64 per cent of the total primary students are in 16.45 per cent of the primary schools concentrated in Ahmedabad and Kheda districts. The number of primary teachers per 10,000 students is the highest in the Central Gujarat region (303 teachers) and in Kheda district (348 teachers). The spatial pattern clearly shows the development of primary education in the Central Gujarat region comprising Gandhinagar, Ahmedabad and Kheda districts. Among all the talukas the ratio of the number of primary teachers per 10,000 students is the highest in Kotda Sangani Mahal (1,295 teachers) and the lowest in Morvi taluka (77 teachers) of Rajkot district.

The teacher-student ratio at various levels of education indicates the possible extent of inter-personal interaction between students and teachers. The optimum size of students at various levels of education may be determined depending upon the desired degree of interaction between the teacher and the taught. By and large, the number of students per teacher decreases with the higher level of education. This trend is partly due to the large number of dropouts and largely due to the typical age pyramid with two-fifths of population below the age-group of 14. The teacher-student ratio (the number of students per teacher) is the highest (45 students) in primary schools of the North Gujarat region and the lowest (33 students) in the Central Gujarat region. Junagadh, Amreli (44 students each) and Banaskantha (43 students) districts have a relatively more number of students per primary teacher than the Dangs–33 students per teacher (Table 4.4). Thus, the ratio gives some indication of the number of students in the classrooms.

Medical Facilities

The availability of medical services in terms of the number of doctors per one lakh population and the availability of hospital beds per 10,000 population have been analysed. These two aspects largely determine the curability of a disease and the health of an individual. The more the number of doctors and the beds per health unit the shorter the waiting time and the more timely medical services to the needy. The number of doctors per one lakh population and the number of hospital beds per 10,000 population is relatively high in the Central Gujarat region (11 doctors and 7 beds) especially in the district of Ahmedabad (16 doctors and 10 beds). The other districts having more number of doctors and hospital beds in relation to their population size include Jamnagar, Surendranagar, Gandhinagar and the Dangs. Among the talukas, Abdasa and Lakhpat talukas of Kachchh district and Sihor (Bhavnagar district), Jamnagar and Wadhwan talukas (Surendranagar district) have a relatively more number of doctors and hospital beds respectively (when compared to their population sizes). Though the absolute number of doctors and hospital beds is greater in the urbanized talukas such as Ahmedabad, Vadodara, Junagadh, Jamnagar, Chorasi (Surat), Rajkot and Bhavnagar these services are inadequate in relation to their population sizes. The number of deaths per 1,000 live persons was more in Ahmedabad district (8 deaths) during 1980 and least in Jamnagar district (4 deaths). Thus, despite the concentration of doctors and other facilities in the Central, South and North Gujarat regions (Ahmedabad, Kheda, Bharuch, Surat, Mehsana and Banaskantha districts) the number of deaths per 1,000 persons has not been controlled. In Jamnagar, Vadodara, Bhavnagar and Sabarkantha the death rate is much lower compared to other parts of the State.

Number of Post Boxes

The greater the number of post offices, post boxes, and the better the accessibility to them the higher the degree of communication between the settlements. The means of communication are as important as personal communication in strengthening social relations and harmonious living. The effective communication through modern means has strengthened better social well-being. The number of post boxes are more per 10,000 population (9 and above post boxes) in the peripheral districts of Gujarat (Sabarkantha, the Dangs, Bharuch, Surat, Kachchh and Panchmahals) as the facilities in such terrain condition are area-based rather than population-based (Table 4.4). Among the micro-regions, the Eastern hilly region, the Kathiawar Coastal and South Gujarat regions show a relatively higher number of post boxes per 10,000 population (8 and above post boxes). Among the talukas Jambughoda (Panchmahals), Manavadar (Junagadh), Valod (Surat), the Dangs (Dangs) and Lakhpat (Kachchh) have more than 30 post boxes per 10,000 population. Rajkot, Ahmedabad City, Borsad and Sinor talukas have two or three post boxes per 10,000 population (Appendix III). This is partly because of the high density of population and the large size of the post boxes. However, the residents in the peripheral urban areas experience the inadequacy and inaccessibility to the post boxes. In a densely populated urban area there is a need for bigger post boxes rather than greater number of post boxes. Adequate care should be taken to identify the most accessible location in terms of maximum population and interaction patterns. Invariably the post boxes are located at the crossroads, major shopping complexes, administrative offices, higher educational and research institutions, centre of the high density residential areas, hotels and pan shops, and the transport terminals (but stands, railway stations, airports, etc.).

Electric Power Supply

The well-being of an individual or a community as a whole depends largely on the availability of electric power. The modern living whether in rural areas or in urban areas heavily depends on new technology which is power-based. Similarly the innovations in various economic sectors use power to a greater extent. Most of the modern sources of comforts and well-being such as television, radio, refrigerator, water-heater, air-conditioning, oven, room-heater, lighting and other domestic appliances consume electric power. Hence, the intensity of electrification by settlements and talukas indicates indirectly the degree of general well-being. By and large, the percentages of electrification of settlements in Gujarat State is fairly high (75 and above) except the Eastern hilly region (only 36). The hilly, forested and tribal inhabited Eastern region has numerous physical constraints for extending the power supply to its sparsely distributed settlements. Kheda is the only district in Gujarat which has achieved 100 per cent electrification of its settlements. The other districts achieving more than 80 per cent electrification of their settlements are Mehsana (94%), Ahmedabad (86%), Bhavnagar (85%), Rajkot (81%), and Sabarkantha (80%). Out of a total of 184 talukas of Gujarat only 29 talukas have achieved 100 per cent electrification (about 16 per cent of the total talukas) in their settlements. Lakhpat (Kachchh), the Dangs, Uchchhal (Surat), Shehera Devgadhbaria, Limkheda, Jhalod (Panchmahals), Nasvadi (Vadodara), Dediapada, Sagbara (Bharuch), Songadh (Surat), Dharmpur (Valsad) talukas have less than 25 per cent of their settlements with electrification (Appendix III). The tribal Development Boards committed to the cause of tribal well-being may intensify their activities and in collaboration with the Electricity Board may provide electric power to all the settlements in the region.

Transport Network – An Indicator of Economic Health

The transport network reflects the economic health at various geographic scales. The density of transport route network possibly indicates the intensity of interaction including the movement of commodities and passengers. The transport arteries have remained instrumental in bringing the towns in close contact with the other parts of the region and the State and also in promoting and strengthening backward and forward linkages of the industrial units located in major towns. Gujarat State had 3.91 per cent (60,497 kms) of the total length of roads in India (15,45,891 kms) including both the surfaced and unsurfaced roads during 1981-82. In the total surfaced network of roads in India (7,31,959 kms), Gujarat had a relatively higher percentage (6.06% i.e. 44,355 kms) during the same period. The density of transport network largely depends on tropographical ease and on the intensity of economic activities. Among the micro-regions of Gujarat, the Central Gujarat (578 kms) and South Gujarat (502 kms) plain regions have more than 500 kms of road network per 1,000 sq km area (Table 4.3). These micro-regions show a high degree of urban and industrial activities including the agricultural development in rural areas. The Kathiawar (north) and Kachchh region has a low density of road network on account of arid, sandy, saline soil, low agricultural productivity and sparsely distributed population and settlements. The inter-district variation in the density of road network ranges from 85 kms of road network per 1,000 sq km area in Kachchh district to 674 kms of network in Valsad district. There are five districts, namely, Valsad (674 kms), Kheda (658 kms), Dangs (581 kms), Panchmahals (527 kms) and Sabarkantha (500 kms) which have more than 500 kms of road network per 1,000 sq km area. The inter-taluka variation in the network pattern shows a wide range. Ahmedabad City taluka including the adjacent Daskroi taluka show the highest density of 4,589 kms road network and the lowest density of 84 kms route network per 1,000 sq km area in Bhachav taluka of Kachchh district. It is generally believed that the density of transport network will be more in talukas having a higher percentage of urban population and concentration. However, there are in all 42 talukas having more than 500 kms of route network per 1,000 sq km area of which 11 talukas do not have any urban population. There are 18 talukas with less than 20 per cent, another 11 talukas with 20–40 per cent and the remaining 2 talukas with above 40 per cent urban population that have a high density of road network. This clearly shows that even the hilly, forested talukas and districts such as the Dangs, Panchmahals, Sabarkantha, Surat and Valsad can show a high density of transport network on account of the zig-zag route path as well as due to their peripheral location in the State linking not only the areas of the State but also the rural and urban settlements of the adjacent States, namely, Maharashtra, Madhya Pradesh and Rajasthan.

Cinema as Mass Recreator

Despite the widespread use of television in Indian society, the Cinema still remains a major recreator especially for the majority of the poor. This media offers maximum recreation at minimum expenditure and forges unity and well-being among diversified social groups. The Indian Cinema has played an instrumental role in bringing social awareness and social transformation since its inception. Therefore, it is pertinent here to understand the nature and intensity of distribution of cinema halls in relation to the population size, economic activities and social composition. The distribution of cinema halls shows a close association with the urban and industrial societies. Among the 19 districts of the State, eight districts, namely, Ahmedabad (69), Surat (61), Kheda (43), Rajkot (39), Vadodara (37), Junagadh (36), Mehsana (31), and Bhavnagar (30) have 30 and more cinema halls (Table 4.4). However, in relation to population size, Surat district shows a relatively more number of cinema halls (2.45 ratio) per one lakh population. Some of the other districts to follow in order are Rajkot, Amreli, Ahmedabad, Junagadh and Bhavnagar. The Dangs district does not have any permanent cinema halls. Gandhinagar, Panchmahals, Sabarkantha and Banaskantha have a less number of cinema halls in relation to their population sizes. Among the micro-regions the South Gujarat region comprising some of the talukas of Valsad, Surat, Bharuch and Vadodara shows more number of cinema halls per one lakh population. The region experiences industrial impact, social heterogeneity and celluloid culture of the Bombay metropolis due to its physical proximity. Out of a total of 184 talukas only five talukas namely Ahmedabad City (55), Chorasi (29), Vadodara (19), Rajkot (14) and Anand (12) have more than 10 cinema halls. In terms of number of cinema halls per one lakh population Umrala, Okhamandal, Palasana, Jafrabad, Bhavnagar, Bardoli and Valod show better position, i.e. more than three cinema halls per one lakh population. Thus it is clear from the analysis that there exist three major centres for the purposes of recreating residents, namely, Ahmedabad, Surat and Rajkot. The other intermediate centres of cinema recreation are Vadodara, Anand, Nadiad, Junagadh, Bhavnagar and Jamnagar.

The Spatial synthesis of the above discussed socio-economic indicators of well-being reveal interesting pattern (Fig. 4.1). Out of a total of 184 talukas nearly 44 talukas show a relatively better level of well-being on the basis of these indicators (Table 4.5). There are eight talukas each in Sabarkantha and Surat districts which show a better level of well-being. They are Modasa (21.20), Bayad (26.26), Malpur (28.73), Vijaynagar Mahal (29.23), Himatnagar (31.96), Bhiloda (35.53), Meghraj (36.74), Idar (37.09) (of Sabarkantha district) and Bardoli (26.56), Kamraj (27.58), Mahuva (30.33), Mangrol (33.49), Uchchhal (33.59), Palasana (34.34), Olpad (35.63), Valod (35.73), Nizar (38.15)(of Surat district). The other talukas of the South Gujarat region include Hansot, Jhagadia, Nanded, Valia (Bharuch), Bansda,

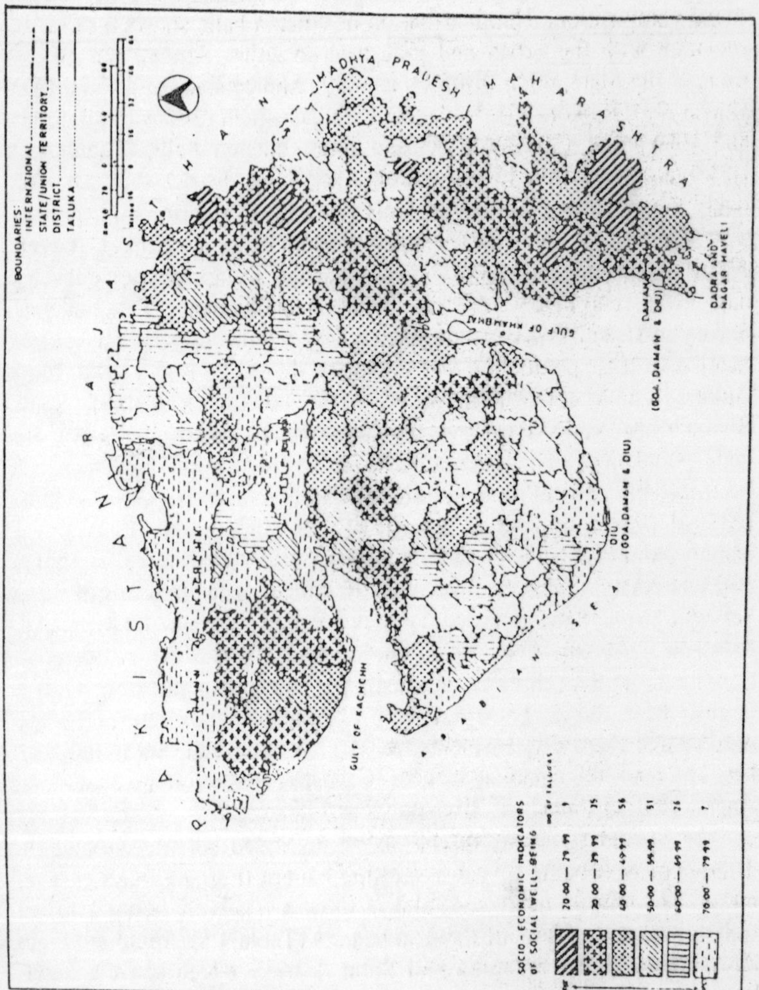

FIG. 4.1: *Patterns of Social Well-being in Gujarat*
(based on socio-economic index value)

Valsad, Pardi and Gandevi (Valsad district). The one taluka district of the Dangs also shows a better level of well-being (part of Eastern hilly region). Similarly, four talukas of Kachchh district, namely, Bhuj, Mundra, Abdasa and Lakhpat, show a higher level of well-being. The well-being of people in the peripheral areas is mainly due to the natural environment and the minimum needs of the tribal societies which are met by the governmental schemes. There are other isolated pockets of higher well-being in Saurashtra such as Jamnagar, Dhrol, Maliya, Wankaner, Sihor, Vallabhipur Mahal. The Central Gujarat plain region with Ahmedabad City taluka as magnet extends to cover Mehmedabad, Nadiad, Khambat, Matar and Balasinor of Kheda district expressing a higher level of well-being by way of socio-economic indicators.

TABLE 4.5: Distribution of Talukas by Socio-economic Index Values

Socio-economic index values	*Number of talukas*	*Level of well-being*
20.00 - 29.99	9	Very high
30.00 - 39.99	35	High
40.00 - 49.99	56	Moderately high
50.00 - 59.99	51	Moderately low
60.00 - 69.99	26	Low
70.00 - 79.99	7	Very low

There are 33 talukas in Gujarat showing a relatively low level of well-being of which 21 are in the Saurashtra region. Another nine talukas are situated in Banaskantha and Mehsana districts of the North Gujarat region. Mangrol, Visavadar, Una, Patan Veraval of Junagadh district, Kalyanpur of Jamnagar district, Jasadan of Rajkot district and Sami taluka of Mehsana district show a very low level of well-being. These areas experience aridity, shifting sandy saline soil (in Little Rann only saline sandy soil) and low productivity resulting into backwardness. These talukas are eligible to receive the required socio-economic inputs from the concerned government departments so that the overall pattern of inequalities is minimized.

The average index value of the socio-economic indicators for the whole of the State (184 talukas) is 48.86 which is almost equal to that of the average of Demographic index value (48.48). Ninety-six talukas out of total of 184 (52% of the total talukas) record more than the average index value for the socio-economic indicators. Like the demographic index value the socio-economic index value too is partial viewpoint to express the level of well-being. The distribution pattern of 21 talukas of the Saurashtra region with a low level of well-being is as follows: Kathiawar (North) and Kachchh (6 talukas), Kathiawar core (8 talukas), and Kathiawar coastal (7 talukas) (Appendix V).

Thus, the present chapter brings out the pattern of well-being by talukas on the basis of certain selected diagnostic socio-economic indicators. The composite index of social well-being is attempted by combining demographic and socio-economic index values in the following section.

5

SPATIAL PATTERNS OF SOCIAL WELL-BEING IN GUJARAT

AN attempt has been made here to synthesize the demographic and socio-economic indicators (index values) under study to identify and analyse the territorial patterns of social well-being in Gujarat. This has been achieved by adopting two different geographic scales and sets of variables. One of the two approaches identifies the in-equalities of well-being by talukas on the basis of State as a unit of study (Fig. 5.2) and the other considers micro-region as a unit and identifies inequalities within (Fig. 5.1) the region. However, the level of social well-being measured for each of the talukas in Gujarat remains tentative and is considerably influenced by the initial selection of variables for the micro-regional analysis and by the final selection of a few indicators of social well-being at the State level.

Intra-Regional Pattern

The seven micro-regions of Gujarat, except the Eastern Hilly region, show conformity of the level of social well-being with the existing social situation. The talukas identified as having a very high level of social well-being within the Eastern Hilly region are in relation to its regional average and are not comparable with the similar talukas of other regions. This kind of exploratory study particularly helps planners and administrators to understand the order in which the various talukas are responding to their socio-economic programmes and inputs of the preceding plan period. By and large, the better social well-being is observed in the Gujarat plains rather than in the Kathiawar Peninsula and the Eastern hilly region where the isolated pockets of social well-being are observed. Among the micro-regions the following is the decreasing order of social well-being: the Central Gujarat plain, the South Gujarat plain, the Kathiawar Core, the Kathiawar (north) and Kachchh, the Kathiawar Coastal, the North Gujarat plain and the

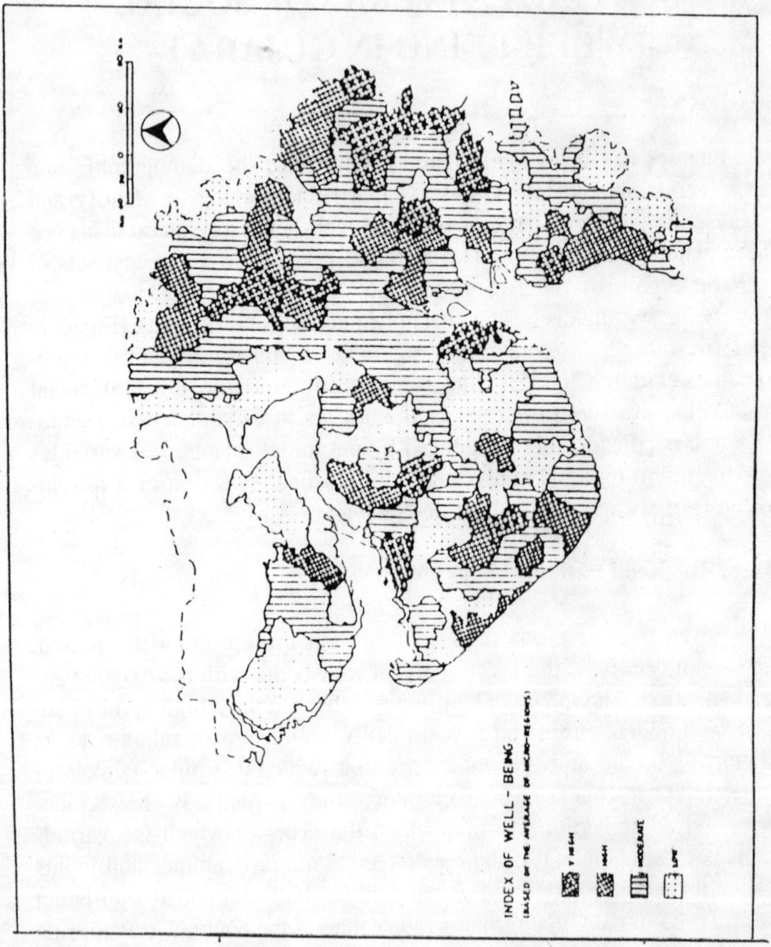

FIG. 5.1: *Social Well-being in Gujarat*
(based on the average of micro-regions)

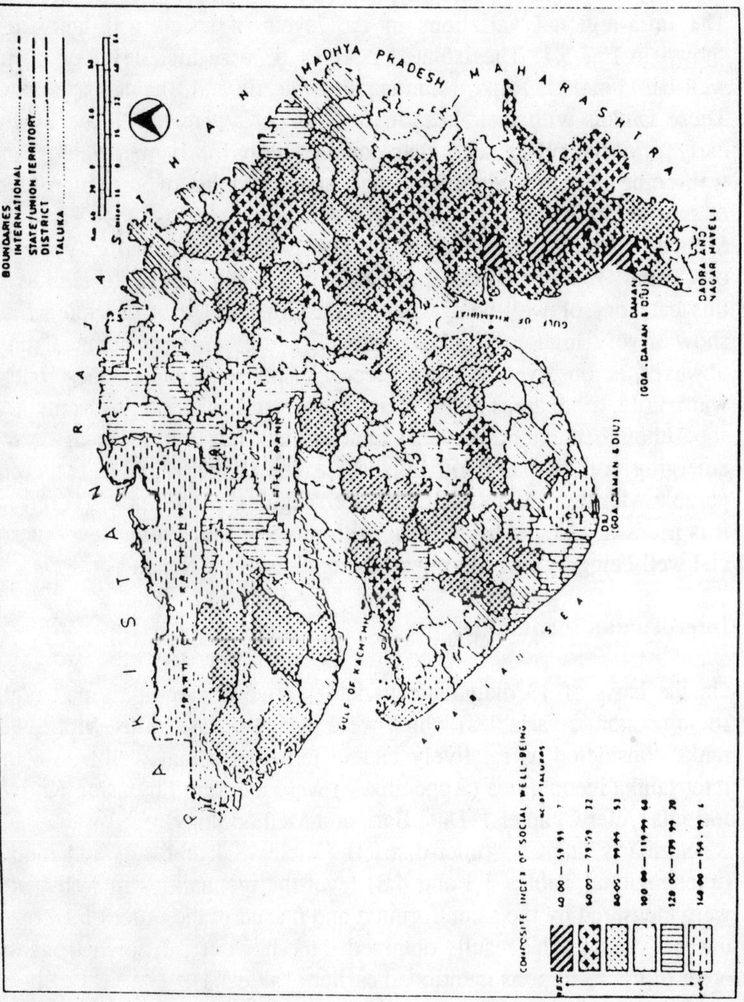

FIG. 5.2: *Patterns of Social Well-being in Gujarat*
(based on composite index values)

Eastern Hilly region. On the basis of the composite score and population size of each of the 184 talukas, four significant levels of social well-being are identified, such as very high (16 talukas), High (32 talukas), Moderate (76 talukas), and Low (60 talukas) (Appendix IV). The intra-regional variations in the levels of social well-being are shown in Fig 5.1. The isolated pockets of very high level of social well-being include Rajkot, Jamnagar, Junagadh and Bhavnagar talukas. These talukas with their headquarters as core were developed by the early princely rulers with due consideration for human values and well-being. Another Significant and extensive area of very high level of social well-being is in North Gujarat covering the Visnagar, Sidhpur, Mehsana, Kalol areas merging with Ahmedabad City region of Central Gujarat region. There are five talukas (out of 16 talukas in this category of well-being) in the Eastern Hilly tribal region which show a very high level of social well-being. This phenomena may, however, be observed as the degree of response shown by these backward, hilly, tribal areas to the various programmes of the government.

Although the intra-regional analysis has its own advantages, the emerging pattern based on talukawise composite value is not comparable with the talukas of other macro-regions of the State. Therefore, it is necessary to identify talukas with more or less similar level of social well-being by considering all the talukas of the State.

Inter-Taluka Inequalities

On the basis of 13 diagnostic variables (five demographic and eight socio-economic variables) which were placed in rank order with lower ranks considered as relatively bad in terms of social well-being, the inter-taluka inequalities (Appendix V) were measured by using Knox's formula (refer Chapter 1, Data Base and Methodology).

Similarly at meso (inter-district) (Tables 3.4 and 4.4) and macro (inter-regional Tables 3.3 and 4.3) level the inequalities in well-being were measured by the same formula and placed in the order of decreasing well-being. The results obtained thus, however, differed from the order of micro-regions mentioned earlier (Table 5.1).

Out of a total of 184 talukas, 39 talukas show a relatively better level of social well-being (very High and High) (Table 5.3). Among them 24 talukas are situated in the South Gujarat region (Valsad-10, Surat-4, Bharuch-4, Vadodara-4, Kheda-1 and Panchmahal-1), 8 talukas in Central Gujarat (Kheda-4, Sabarkantha-3, Ahmedabad-1), 5 talukas in the Eastern hilly region (Surat-2, Bharuch-2, Dangs-1) and one each in the Kathiawar (north) and Kachchh (Jamnagar) and Kathiawar Core (in Amreli district) regions. The low and very low level of social well-being is observed in 24 talukas with major spatial concentrations in the North Gujarat region (9 talukas), the Kathiawar Coastal (6 talukas) and Kathiawar (North) and Kachchh regions (5 talukas) (Fig. 5.2). The Kathiawar core and the Eastern hilly region have three and one taluka respectively. Among the districts, Banaskantha has more talukas (7) with a low level of well-being followed by Junagadh (6), Jamnagar, Amreli, Kachchh, Mehsana (2 talukas each) and Rajkot, Surendranagar and Panchmahal (1 taluka each). Refer Appendix V for talukawise level of well-being. Sami taluka of Mehsana district has been observed to have the lowest level of well-being in Gujarat.

TABLE 5.1: Demographic, Socio-economic and Composite Index of Well-being (micro-regions arranged in the decreasing order of well-being)

	Micro regions	*Composite index of well-being*	*Index of well-being based on demographic indicators*	*Index of well-being based on socio-economic indicators*
1.	South Gujarat Plain	68.93	24.29	44.64
2.	Central Gujarat Plain	75.36	34.29	41.07
3.	Kathiawar core	121.79	62.86	58.93
4.	Kathiawar (north) and Kachchh	123.57	70.00	53.57
5.	Eastern hilly region	131.07	68.57	62.50
6.	Kathiawar Coastal	134.29	62.86	71.43
7.	North Gujarat Plain	145.00	77.14	67.86

The inter-district inequalities in well-being are shown in Table 5.2. The data also reveal that the districts of Surat and Ahmedabad have a high degree of well-being and Junagadh and Banaskantha a low degree of well-being. While analysing the inter-taluka variation it has been observed that the majority of talukas in these two districts showed extreme levels of well-being respectively. Surat, Ahmedabad, Valsad and Vadodara districts showed a very high level of well-being, followed by Rajkot, Dangs, Kheda with a high level, Gandhinagar, Sabarkantha, Bharuch, Bhavnagar, Jamnagar, Surendranagar with a moderate level of well-being and Kachchh, Panchmahal, Amreli, Mehsana, Junagadh, Banaskantha with a low level of well-being.

TABLE 5.2: Demographic, Socio-economic and Composite Index of Well-being (districts arranged in the decreasing order of well-being)

	Districts	*Composite index of well-being*	*Index of well-being based on demographic indicators*	*Index of well-being based on socio-economic indicators*
	1	*2*	*3*	*4*
1.	Surat	64.21	16.84	47.37
2.	Ahmedabad	67.90	23.16	44.74
3.	Valsad	73.81	28.42	45.39
4.	Vadodara	77.89	25.26	52.63
5.	Rajkot	84.21	31.58	52.63
6.	Dangs	86.46	63.16	23.30
7.	Kheda	95.79	48.42	47.37
8.	Gandhinagar	105.13	50.53	54.60
9.	Sabarkantha	105.79	71.58	34.21
10.	Bharuch	106.05	49.47	56.58
11.	Bhavnagar	107.63	53.68	53.95
12.	Jamnagar	108.29	58.95	49.34
13.	Surendranagar	113.16	68.42	44.74
14.	Kachchh	122.50	70.53	51.97
15.	Panchmahals	123.55	55.79	67.76
16.	Amreli	125.53	66.32	59.21
17.	Mehsana	128.42	60.00	68.42
18.	Junagadh	131.19	62.11	69.08
19.	Banaskantha	148.16	81.05	67.11

TABLE 5.3: Distribution of Talukas by Composite Index Value

Composite index values	*Number of talukas*	*Level of social well-being*
40.00 - 59.99	7	Very high
60.00 - 79.99	32	High
80.00 - 99.99	53	Moderately high
100.00 - 119.99	68	Moderately low
120.00 - 139.99	20	Low
140.00 - 159.99	4	Very low

Among the seven micro-regions, South Gujarat and Central Gujarat region show a high index of well-being. Kathiawar Core and Kathiawar (north) and Kachchh record a moderate level and the remaining micro-regions – Eastern Hilly region, Kathiawar Core and North Gujarat region – record a low level of well-being (Table 5.1). The six talukas of Junagadh (5) and Amreli (1) and nine talukas of Banaskantha (7) and Mehsana (2) districts have mainly contributed to the low level of well-being in the Kathiawar Coastal and North Gujarat regions respectively. The micro-regions with a low level of well-being need to be strengthened by way of demographic and socio-economic inputs. The Kathiawar Coastal region particularly needs more of socio-economic inputs (chosen indicators under study) rather than improving the demographic situation (by way of qualitative population resource).

6

SOME OBSERVATIONS AND CONCLUSIONS

THE present study has made an attempt to reveal more explicitly the patterns of inequalities in the levels of well-being in Gujarat on the basis of certain selective demographic and socio-economic indicators. However, one may expect marginal changes in the order of talukas presented in the study depending upon the selection of indicators. The study highlights the disparities in the levels of well-being by choosing three different sets of geographic scales, namely, taluka, district and micro-region. At all the levels, the core-peripheral contrasts in well-being emerge with the exception of Nizar, Uchchhal, Dangs talukas of the Eastern hilly region. These talukas show a high level of well-being perhaps due to the hydel power projects across Tapi river and the tribal development schemes.

The core-peripheral contrast in the levels of well-being in Gujarat appears not because of the theoretical advantages of economic development in the core areas but because of the topographical, climatic, soil and water conditions along with the aspects of historical development. The peripheral areas of Gujarat State are at a disadvantage as far as the topographical, climatic and soil conditions are concerned. Therefore, the distribution of population resource is meager in these areas (Eastern Hilly areas, North Gujarat, North Kathiawar and Kachchh and Kathiawar Coastal regions). However, the greater socio-economic interaction between the Ahmedabad and Bombay regions has led to the spatial diffusion of innovations and the aspects of development and well-being. The South Gujarat region being situated adjacent to the Bombay metropolitan region receives greater and continuous impetus of modernization and development. The concentration of the Parsee community in this region has promoted the developmental aspects of the society. The progressive community of farmers of Kheda district has taken advantage of the fertile soil and the irrigational facilities available and has developed agriculture. The leading farmers with their sense of commitment and dedication to the cause of social welfare have developed educational institutions in and around Kheda district.

The inequalities in the levels of well-being become more accentuated on account of outmigration of economically productive labour force from the hill talukas of Panchmahal, Sabarkantha, Banaskantha, Bharuch, Dangs and some talukas of Kachchh district (all peripheral talukas of the State). Although the hard-earned money of the migrants is remitted to their place of origin, it becomes inadequate to make any significant impact on the level of living and well-being. Like the distribution of urban population, the general pattern of well-being is confined to certain spatial pockets and also to certain social classes.

One important advantage of the generalization of well-being made here at the taluka level is that of comparison of its ranking order. The most popularly regarded talukas (Ahmedabad city, Vadodara, Nadiad, Anand, Rajkot, Bhavnagar, Chorasi and Jamnagar) with their well-known urban centres, industrial centres and commercial centres do not figure in the top-ranking talukas in terms of the index of well-being although they show a high level of well-being. This clearly shows that the over-all level of well-being in the talukas with larger cities is not all that better. This phenomena is largely due to the inadequate housing, schooling, medical, transport, communication and other recreational facilities in relation to the population size. Therefore, invariably the quality of living and levels of living are much higher in less urbanized areas on account of the low cost of living and better social harmony.

On the other hand the spatial inequalities in social well-being may perhaps be better understood in the light of the variation in the social aspirations, social needs, social values and social attainments as they are controlled by the general pattern of education, occupation, religion, caste, family size, age-sex composition, rural-urban composition and income. However, these controlling factors are ever changing and hence the change in social needs, values, attainments and well-being. The phenomena of social well-being is most dynamic at all levels of society. Although the governmental efforts at the national, State and local levels aim at minimizing the inequalities in the socio-economic and developmental aspects between the societies (caste/religion based), regions (urban, rural and tribal) and the classes (economic or social) the changes in the levels of social well-being do not show any progress in the direction of attaining the goal. Now the question is whether it is possible to achieve the order of uniformity all over or does it remain as a mirage?

APPENDICES

APPENDIX I

MICRO-REGIONAL FRAMEWORK AND ADMINISTRATIVE DIVISIONS

Micro-region	*District*	*Name of Taluka*
1	*2*	*3*
I. **North Kathiwar and Kachchh** (Number of Talukas 35)	1. Jamnagar (Entire)	(i) Jamnagar (ii) Lalpur (iii) Jamjodhpur (iv) Bhanaud (v) Kalyanpur (vi) Okhamandal (vii) Khambhata (viii) Jodiya (ix) Dhrol (x) Kalavad
	2. Rajkot (Partly)	(i) Rajkot (ii) Lodhika Mahal (iii) Paddhari (iv) Morvi (v) Maliya (vi) Wankaner
	3. Surendra-nagar (Entire)	(i) Wadhawan (ii) Limbdi (iii) Sayla (iv) Chotila (v) Muli (vi) Halvad (vii) Dhrangadhra (viii) Dasada (ix) Lakhtar
	4. Kachchh (Entire)	(i) Bhuj (ii) Mundra (iii) Mandvi (iv) Abdasa (v) Lakhpat (vi) Nakhatrana (vii) Rapar(viii) Bhachav (ix) Anjar
	5. Ahmedabad (Partly)	(i) Viramgam
II. **Kathiawar Core** (Number of Talukas 25)	1. Rajkot (Partly)	(i) Kotda Sangani Mahal (ii) Gondal (iii) Jetpur (iv) Dhoraji (v) Upleta (vi) Jamkandorna (vii) Jasdan
	2. Bhavnagar (Partly)	(i) Kundla (ii) Gariadhar (iii) Palitana (iv) Sihor (v) Umrala (vi) Gadhada (vii) Botad (viii) Villabhipura Mahal
	3. Amreli (Partly)	(i) Amreli (ii) Dhari (iii) Khambha Mahal (iv) Kunkavav Vadia (v) Babra (vi) Lathi (vii) Lilia Mahal
	4. Junagadh (Partly)	(i) Junagadh (ii) Bhesan Mahal (iii) Visavadar

(Appendix I Contd.)

1	*2*	*3*
III. **Kathiawar Coastal** (Number of Talukas 19)	1. Bhavnagar (Partly)	(i) Bhavnagar (ii) Ghogho Mahal (iii) Talaja (iv) Mahuva
	2. Amreli (Partly)	(i) Rajula (ii) Jafrabad (iii) Kodinar
	3. Junagadh (Partly)	(i) Mendarda (ii) Talala (iii) Patan-Veraval (iv) Malia (v) Mangrol (vi) Keshod (vii) Porbandar (viii) Ranavav (ix) Kutiyana (x) Manavadar (xi) Vanthali (xii) Una
IV. **North Gujarat** (Number of Talukas 22)	1. Banaskantha (Entire)	(i) Palanpur (ii) Vadgam (iii) Deesa (iv) Kankrej (v) Radhanpur (vi) Santalpur (vii) Deodar (viii) Vav (ix) Tharad (x) Dhanera (xi) Danta
	2. Mehsana (Entire)	(i) Mehsana (ii) Kadi (iii) Chansma (iv) Sami (v) Harij (vi) Patan (vii) Siddhpur (ix) Visnagar (x) Vijapur (xi) Kalol
V. **Central Gujarat** (Number of Talukas 24)	1. Sabarkantha (Entire)	(i) Himatnagar (ii) Prantij (iii) Idar (iv) Khedbrahma (v) Vijaynagar Mahal (vi) Bhiloda (vii) Meghraj (viii) Modasa (ix) Malpur (x) Bayad
	2. Gandhinagar (Entire)	(i) Gandhinagar
	3. Ahmedabad (Partly)	(i) Ahmedabad City (ii) Daskroi (iii) Dholka (iv) Sanand (v) Dehgam
	4. Kheda (Partly)	(i) Mehemdabad (ii) Nadiad (iii) Anand (iv) Petlad (v) Matar (vi) Kapadvanj (vii) Balasinor (viii) Thasra
VI. **South Gujarat** (Number of Talukas 38)	1. Ahmedabad (Partly)	(i) Dhandhuka
	2. Kheda (Partly)	(i) Borsad (ii) Khambat
	3. Panch-mahals (Partly)	(i) Halol (ii) Jambughoda

(Appendix I Contd.)

1	*2*	*3*
	4. Vadodara (Partly)	(i) Vadodara (ii) Karjan (iii) Padra (iv) Savli (v) Vaghodia (vi) Dabhoi (vii) Sankheda (viii) Tilakwada Mahal (ix) Sinor
	5. Bharuch (Partly)	(i) Bharuch (ii) Ankleshwar (iii) Hansot (iv) Vagra (v) Jambusar (vi) Amod
	6. Surat (Partly)	(i) Chorasi (ii) Olpad (iii) Kamrej (iv) Mangrol (v) Mandvi (vi) Vyara (vii) Valod (viii) Bardoli (ix) Mahuva (x) Palsana
	7. Valsad (Entire)	(i) Valsad (ii) Pardi (iii) Umbergaon (iv) Gandevi (v) Navsari (iv) Chikhli (vii) Bansda (viii) Dharmpur
VII. **Eastern Hilly Region** (Number of Talukas 21)	1. Panch-mahal (Partly)	(i) Godhra (ii) Kalol (iii) Shehera (iv) Lunawada (v) Santrampur (vi) Jhalod (vii) Dohad (viii) Limkheda (ix) Devgadhbaria
	2. Vadodara (Partly)	(i) Jetpurpavi (ii) Chota Udepur (iii) Nasvadi
	3. Bharuch (Partly)	(i) Jhagadia (ii) Nandod (iii) Dediapada (iv) Sagbara (v) Valia
	4. Surat (Partly)	(i) Songadh (ii) Uchchhal (iii) Nizar
	5. Dangs (Entire)	(i) Dangs

Total number of districts in Gujarat - 19 and talukas - 184.
District-wise number of talukas:

(1) Jamnagar-10
(2) Rajkot - 13
(3) Surendranagar - 9
(4) Kachchh - 9
(5) Bhavnagar - 12
(6) Amreli - 10
(7) Junagadh - 15
(8) Banaskantha - 11
(9) Mehsana - 11
(10) Sabarkantha - 10
(11) Gandhinagar - 1
(12) Ahmedabad - 7
(13) Kheda - 10
(14) Panchmahals - 11
(15) Vadodara - 12
(16) Bharuch - 11
(17) Surat - 13
(18) Valsad - 8
(19) Dangs - 1

APPENDIX II

TALUKAWISE SOME IMPORTANT DEMOGRAPHIC INDICATORS

Taluka	*Density of population (per sq km)*	*Percentage of urban population*	*Percentage of female literacy*	*Percentage of male workers*	*Percentage of female workers*
1	*2*	*3*	*4*	*5*	*6*
1. Jamnagar	385 (23.5)	70.7 (6)	39.2 (29)	57.5 (16)	7.1 (158)
2. Lalpur	82 (167.5)	10.9 (117)	24.6 (115)	53.7 (57)	12.4 (74)
3. Jamjodhpur	101 (156)	17.6 (78)	34.4 (54)	53.6 (58)	14.4 (50.5)
4. Bhanvad	126 (139)	16.7 (84)	24.1 (121)	52.9 (80)	12.1 (79)
5. Kalyanpur	85 (164.5)	-	14.0 (158)	53.7 (54)	5.1 (180)
6. Okhamandal	131 (133.5)	60.3 (9)	29.4 (77.5)	50.3 (150)	5.5 (178)
7. Khambata	122 (141)	29.7 (32)	18.7 (144)	52.2 (95.5)	6.0 (173)
8. Jodiya	96 (159.5)	13.6 (104)	32.0 (64)	50.0 (158)	11.2 (86)
9. Dhrol	102 (153.5)	25.4 (43)	29.4 (79)	50.2 (154)	8.8 (120)
10. Kalavad	102 (153.5)	13.2 (107)	31.3 (67)	51.6 (114.5)	14.1 (53.5)
11. Rajkot	531 (10)	78.6 (4)	49.8 (4.5)	49.7 (162)	5.9 (175)
12. Kotda-Sangani	115 (146)	-	31.1 (68)	51.4 (120)	19.7 (25)
13. Gondal	178 (98.5)	31.4 (29)	43.0 (16)	50.2 (156)	11.1 (87)
14. Jetpur	256 (51)	42.6 (15)	38.7 (32)	51.3 (122.5)	8.0 (135)
15. Dhoraji	313 (36)	51.2 (11)	44.4 (11)	51.5 (117)	12.4 (72)
16. Upleta	215 (74)	32.2 (27)	44.1 (12)	52.4 (89)	16.3 (37)
17. Jamkandorna	114 (147.5)	-	35.5 (47)	52.3 (94)	13.7 (59)

(Appendix II Contd.)

1	*2*	*3*	*4*	*5*	*6*
18. Lodhika Mahal	101 (156)	-	20.3 (81)	53.2 (69)	16.3 (37)
19. Paddhari	101 (156)	9.2 (125)	29.3 (32)	52.0 (105)	19.0 (26)
20. Morvi	144 (125)	32.7 (26)	34.6 (52)	50.3 (151)	7.9 (138)
21. Maliya	82 (167.5)	-	24.8 (113)	50.9 (137)	9.2 (113)
22. Wankaner	110 (149)	26.3 (39)	24.5 (116)	53.3 (66)	16.0 (43)
23. Jasdan	127 (137)	16.2 (89)	19.1 (140)	52.2 (98)	13.1 (64)
24. Wadhwan	259 (48.5)	63.2 (7)	39.9 (27)	49.3 (169)	8.8 (118.5)
25. Limbdi	104 (152)	17.3 (80)	24.3 (120)	50.8 (141)	14.1 (56)
26. Sayla	63 (177)	17.0 (82)	13.4 (161)	53.6 (59)	13.0 (65)
27. Chotila	93 (161.5)	29.0 (33)	15.9 (154)	51.3 (125)	15.0 (47)
28. Muli	73 (173)	-	18.3 (145)	53.0 (73)	16.0 (42)
29. Halvad	74 (171.5)	14.6 (99)	20.5 (134)	54.0 (49)	14.1 (57)
30. Dhrangdhra	100 (158)	37.4 (20)	29.1 (35)	53.4 (64)	10.7 (90)
31. Dasada	84 (166)	15.4 (95)	21.3 (127)	52.1 (102.5)	10.4 (96.5)
32. Lakhtar	76 (170)	19.4 (71)	25.7 (108)	50.5 (148)	10.1 (99)
33. Bhavnagar	278 (44)	77.8 (5)	43.2 (14)	49.4 (166)	5.9 (174)
34. Ghogho Mahal	145 (123)	13.9 (103)	21.0 (132)	53.0 (74)	16.9 (33)
35. Talaja	210 (77.5)	8.1 (132)	15.1 (155)	52.9 (79)	18.1 (29)
36. Mahuva	216 (73)	21.2 (63)	18.2 (146)	51.4 (121)	16.2 (40.5)
37. Kundla	172 (105)	24.7 (48)	30.7 (69)	51.1 (130)	11.6 (83)
38. Gariadhar	184 (95)	17.1 (81)	28.1 (89.5)	53.0 (75)	11.7 (82)
39. Palitana	190 (90)	24.7 (48)	22.8 (123)	52.2 (95.5)	12.7 (67)
40. Sohor	184 (95)	22.7 (56)	20.4 (105)	51.2 (128.5)	10.7 (91)
41. Umrala	178 (98.5)	18.7 (72)	27.5 (97)	54.2 (47)	12.5 (69)
42. Gadhada	138 (130)	13.0 (109)	24.3 (119)	53.3 (68)	13.4 (62)

(Appendix II Contd.)

1	*2*	*3*	*4*	*5*	*6*
43. Botad	177 (100)	43.1 (14)	27.9 (92)	52.7 (82.5)	12.3 (75)
44. Valabhipur Mahal	107 (151)	17.3 (79)	29.3 (80)	54.3 (46)	16.6 (34)
45. Amreli	210 (76.5)	33.1 (24)	46.9 (8)	48.9 (172)	10.3 (98)
46. Dhari	116 (144.5)	23.5 (52)	38.6 (33)	51.0 (133)	10.4 (95)
47. Khambha Mahal	114 (147.5)	-	33.0 (61)	51.2 (127)	14.7 (48)
48. Rajula	156 (119)	20.4 (68)	17.9 (147)	52.6 (86)	12.3 (77)
49. Jafrabad	166 (110)	22.9 (55)	11.7 (166)	52.3 (92)	8.7 (122)
50. Kodinar	259 (48.5)	15.6 (93)	24.9 (112)	50.0 (157)	7.6 (145)
51. Kunkavav Vadia	182 (97)	21.4 (62)	37.2 (37)	49.9 (159)	9.5 (106)
52. Babra	116 (144.5)	13.3 (106)	26.1 (107)	52.4 (90)	10.0 (102)
53. Lathi	159 (116)	25.0 (46)	33.3 (57)	53.0 (72)	15.1 (46)
54. Lilia Mahal	136 (131.5)	-	35.4 (48)	52.5 (88)	8.4 (132)
55. Junagadh	360 (30)	56.0 (10)	44.5 (10)	48.3 (175)	7.2 (155)
56. Mandarda Mahal	160 (114.5)	18.5 (74)	33.2 (58)	51.3 (122.5)	9.6 (104)
57. Talala	96 (159.5)	12.0 (111)	29.4 (76)	49.4 (165)	9.2 (112)
58. Patan Veraval	379 (25.5)	40.4 (16)	20.1 (136)	50.4 (149)	9.3 (111)
59. Malia	206 (81)	9.0 (127.5)	23.5 (122)	50.3 (152)	10.1 (100)
60. Mangrol	251 (53.5)	26.5 (38)	19.6 (138)	46.9 (181)	8.9 (117)
61. Keshod	246 (56)	23.4 (54)	35.2 (49)	48.0 (177)	7.6 (147)
62. Porbandar	239 (64)	49.2 (12)	38.0 (35)	49.4 (167)	5.7 (177)
63. Ranavav	128 (136)	37.9 (17)	27.9 (93)	51.9 (108)	9.5 (107)
64. Kutiyana	141 (129)	22.5 (57)	28.1 (91)	51.1 (132)	9.0 (116)
65. Manavadar	212 (76)	29.8 (31)	41.4 (21)	47.9 (178)	8.2 (133)
66. Vanthali	244 (60)	29.0 (34)	17.3 (150)	49.4 (164)	11.4 (85)
67. Bhesan Mahal	152 (121)		28.1 (89.5)	49.8 (161)	12.1 (78)

(Appendix II Contd.)

1	2	3	4	5	6
68. Visavadar	131	11.7	30.5	50.7	10.7
	(133.5)	(112)	(70)	(142)	(92.5)
69. Una	142	16.9	16.4	50.6	9.4
	(128)	(83)	(153)	(143)	(108)
70. Bhuj	48	37.7	31.5	51.7	12.9
	(181)	(18)	(66)	(113)	(66)
71. Mundra	71	16.1	29.8	50.6	11.9
	(175)	(91)	(74)	(146)	(81)
72. Mandvi	93	24.7	34.8	48.5	12.4
	(161.5)	(48)	(51)	(174)	(73)
73. Absada	33	9.0	24.5	53.7	15.3
	(183)	(127.5)	(117)	(55)	(45)
74. Lakhpat	15	-	17.4	57.8	14.1
	(184)		(149)	(12)	(52)
75. Nakhatrana	53	-	34.5	46.6	16.3
	(179)		(53)	(182)	(39)
76. Rapar	45	7.8	9.0	56.1	6.7
	(182)	(133)	(175)	(30)	(165)
77. Bhachav	50	13.9	13.9	52.1	12.5
	(180)	(102)	(159)	(104)	(70)
78. Anjar	146	62.2	31.5	51.9	8.5
	(122)	(8)	(65)	(110)	(126.5)
79. Palanpur	209	22.1	20.3	51.3	7.8
	(79)	(58)	(135)	(126)	(140)
80. Vadgam	246	-	21.5	50.2	10.8
	(56)		(128)	(155)	(88)
81. Deesa	176	16.0	9.9	51.6	5.8
	(101.5)	(92)	(173)	(114.5	(170)
82. Kankrej	185	-	8.8	53.6	7.0
	(93)		(176)	(60.5)	(160)
83. Radhanpur	136	26.0	14.2	52.8	7.4
	(131.5)	(41)	(157)	(81)	(150.5)
84. Santalpur	54		7.3	53.8	6.2
	(178)		(179)	(53)	(170)
85. Deodar	143	-	6.9	53.6	6.9
	(127)		(180)	(60.5)	(161)
86. Vav	72	-	4.3	55.6	3.9
	(174)		(184)	(34)	(182)
87. Tharad	109	9.0	4.7	52.9	3.6
	(150)	(126)	(183)	(78)	(183)
88. Dhanera	120	-	6.2	54.0	7.7
	(142)		(181)	(48)	(143)
89. Danta	118	-	9.8	52.3	6.3
	(143)		(174)	(93)	(169)
90. Himatnagar	244	21.2	34.3	50.6	12.5
	(60)	(64)	(55)	(147)	(71)
91. Prantij	263	14.3	29.4	50.9	10.7
	(45)	(100)	(77.5)	(140)	(92.5)
92. Idar	213	8.1	29.7	51.2	10.6
	(75)	(131)	(75)	(128.5)	(94)

(Appendix II Contd.)

1	*2*	*3*	*4*	*5*	*6*
93. Khedbrahma	161	10.9	12.8	53.1	8.1
	(113)	(166)	(164)	(71)	(134)
94. Vijaynagar	129	-	19.9	42.7	3.4
Mahal	(135)		(137)	(184)	(184)
95. Bhiloda	202	-	22.1	45.8	7.3
	(85)		(126)	(183)	(154)
96. Meghraj	168	7.1	13.1	51.9	5.3
	(108)	(136)	(162)	(107)	(179)
97. Modasa	223	16.7	29.3	51.0	10.0
	(70)	(85)	(83)	(136)	(101)
98. Malpur	176	7.3	19.5	53.3	10.4
	(101.5)	(135)	(139)	(67)	(96.5)
99. Bayad	222	-	25.0	53.7	9.5
	(71)		(110)	(56)	(105)
100. Mehsana	387	23.8	40.2	48.0	8.5
	(22)	(51)	(24)	(176)	(130)
101. Kadi	261	16.3	36.8	49.2	8.6
	(46)	(87)	(41)	(171)	(124)
102. Chanasma	234	11.0	35.6	47.3	6.3
	(67)	(115)	(46)	(180)	(168)
103. Sami	85	-	10.9	50.6	7.1
	(164.5)		(170)	(145)	(159)
104. Harij	170	21.6	17.2	52.1	7.6
	(107)	(61)	(152)	(101)	(145)
105. Patan	288	26.1	28.1	50.3	8.8
	(41)	(40)	(88)	(153)	(118.5
106. Sidhpur	425	31.4	41.2	48.7	7.4
	(15)	(30)	(22)	(173)	(153)
107. Kheralu	258	16.2	27.4	51.5	9.9
	(50)	(90)	(98)	(118)	(103)
108. Visnagar	404	23.5	44.0	49.9	11.10
	(19)	(53)	(13)	(160)	(84)
109. Vijapur	385	10.9	41.4	47.7	9.0
	(23.5)	(118)	(20)	(179)	(115)
110. Kalol	452	31.8	37.4	49.3	6.4
	(13)	(28)	(36)	(168)	(167)
111. Gandhinagar	445	21.6	38.8	49.2	7.7
	(14)	(60)	(31)	(170)	(142)
112. Ahmedabad	7966	99.1	55.0	49.6	4.7
City taluka	(1)	(1)	(2)	(163)	(181)
113. Daskroi	398	20.6	33.2	53.0	8.5
	(20)	(66)	(59)	(76)	(126.5)
114. Dholka	154	25.2	30.0	52.2	16.4
	(120)	(45)	(72)	(99)	(35)
115. Dhandhuka	80	20.8	27.1	52.7	16.2
	(169)	(65)	(100)	(82.5)	(40.5)
116. Sanand	174	16.4	22.6	51.8	9.3
	(104)	(86)	(125)	(111)	(109.5)
117. Viramgam	158	21.6	26.7	51.0	9.3
	(117.5)	(59)	(103)	(134.5)	(109.5)

(Appendix II Contd.)

1	2	3	4	5	6
118. Dehgam	296 (37)	13.5 (105)	27.7 (96)	50.9 (138.5)	8.0 (136.5)
119. Mehmedabad	424 (16)	19.4 (70)	28.3 (87)	52.1 (100)	8.7 (121)
120. Nadiad	713 (5)	32.7 (25)	40.1 (26)	50.6 (141)	6.8 (163)
121. Anand	758 (4)	25.8 (42)	42.1 (17)	51.5 (116)	8.0 (136.5)
122. Petlad	645 (7)	20.3 (69)	40.1 (25)	52.7 (84)	7.9 (139)
123. Borsad	611 (8)	10.4 (120)	32.1 (66)	53.5 (62)	7.8 (141)
124. Khambhat	221 (72)	28.9 (35)	35.0 (50)	53.5 (63)	6.8 (164)
125. Matar	279 (43)	-	29.9 (73)	52.1 (102.5)	7.4 (148.5)
126. Kapadvanj	324 (35)	11.0 (114)	28.9 (86)	51.8 (112)	6.1 (171)
127. Balasinor	292 (39.5)	14.0 (101)	21.4 (129)	51.0 (134.5)	7.2 (156)
128. Thasra	356 (31)	18.5 (75)	26.7 (104)	53.4 (65)	6.1 (172)
129. Godhra	336 (33)	25.2 (44)	20.7 (133)	52.4 (91)	7.2 (157)
130. Kalol	364 (29)	9.9 (123)	25.1 (109)	53.9 (50)	7.4 (150.5)
131. Halol	253 (52)	17.9 (77)	19.0 (142)	57.3 (18.5)	16.3 (37)
132. Shehera	246 (56)	-	10.3 (172)	54.5 (45)	8.5 (129)
133. Lunawada	245 (58)	10.6 (120)	22.6 (124)	52.9 (77)	12.3 (76)
134. Santrampur	236 (65.5)	3.4 (141)	10.5 (171)	52.0 (106)	6.8 (162)
135. Jhaloda	251 (53.5)	-	11.0 (169)	52.7 (85)	15.8 (44)
136. Dohad	339 (32)	27.8 (36)	17.7 (148)	51.3 (124)	7.4 148.5)
137. Limkheda	199 (87.5)	-	5.5 (182)	55.5 (37)	7.4 (152)
138. Devgadbaria	240 (63)	5.8 (140)	8.6 (177)	57.0 (10)	30.5 (7)
139. Jambughoda	202 (85)	-	13.4 (160)	55.3 (39)	9.1 (114)
140. Vadodara	1459 (3)	79.6 (3)	55.3 (1)	51.5 (119)	6.6 (166)
141. Karjan	228 (68)	11.3 (113)	37.0 (39)	57.2 (20)	17.7 (30)
142. Padra	377 (27)	15.2 (96)	36.9 (40)	55.6 (36)	8.4 (131)

(Appendix II Contd.)

1	*2*	*3*	*4*	*5*	*6*
143. Savli	244	-	26.9	54.8	12.5
	(60)		(101)	(43)	(68)
144. Vaghodia	186	9.0	27.8	57.6	14.1
	(92)	(129)	(94)	(15)	(53.5)
145. Dabhoi	260	27.0	41.5	55.6	11.9
	(47)	(37)	(18.5)	(35)	(80)
146. Sankheda	207	15.5	30.5	57.5	17.4
	(81)	(94)	(71)	(17)	(31)
147. Jetpur Pavi	208	3.1	11.1	57.7	14.2
	(80)	(142)	(168)	(13)	(49)
148. Chhota	175	10.0	8.3	55.0	10.8
Udaipur	(103)	(122)	(178)	(40)	(89)
149. Nasvadi	167	-	14.8	55.7	8.6
	(109)		(156)	(33)	(125)
150. Tilakwada	199	-	27.7	54.8	7.6
Mahal	(87.5)		(96)	(42)	(145)
151. Sinor	236	12.6	40.4	57.7	21.4
	(65.5)	(110)	(23)	(14)	(22)
152. Bharuch	414	45.5	46.4	52.2	14.1
	(17)	(13)	(9)	(97)	(55)
153. Ankleshwar	292	34.0	39.6	56.2	18.6
	(39.5)	(23)	(28)	(28)	(27)
154. Hansot	126	15.1	41.5	55.8	20.4
	(139)	(97.5)	(18.5)	(32)	(23)
155. Vagra	74	-	36.5	54.5	13.5
	(171.5)		(43)	(44)	(60.5)
156. Jambusar	144	17.9	32.6	52.6	8.7
	(125)	(76)	(62)	(87)	(123)
157. Amod	171	15.1	38.1	56.6	19.8
	(106)	(97.5)	(34)	(26)	(24)
158. Jhagadia	158	-	26.9	57.8	23.1
	(117.5)		(102)	(11)	(18)
159. Nandod	160	16.3	29.2	54.9	14.1
	(114.5)	(88)	(84)	(41)	(50.5)
160. Dediapada	86	-	11.5	56.0	40.1
	(163)		(167)	(30)	(1)
161. Sagbara	164	-	17.2	57.1	13.3
	(111)		(151)	(22)	(63)
162. Valia	184	-	25.0	56.6	33.0
	(95)		(111)	(25)	(3)
163. Chorasi	1839	85.5	49.8	57.0	8.5
	(2)	(2)	(4.5)	(24)	(128)
164. Olpad	198	10.4	43.2	55.4	24.3
	(89)	(121)	(15)	(38)	(14.5)
165. Kamraj	284	9.9	36.6	58.4	29.9
	(42)	(124)	(42)	(7)	(8)
166. Mangrol	202	6.3	27.2	57.0	31.6
	(85)	(137)	(199)	(23)	(6)
167. Mandvi	188	8.8	24.7	59.6	23.9
	(91)	(130)	(114)	(2.5)	(16)

(Appendix II Contd.)

1	*2*	*3*	*4*	*5*	*6*
168. Songadh	126	18.5	18.8	58.5	28.7
	(139)	(73)	(143)	(6)	(9)
169. Uchchhal	162	-	11.8	60.5	26.7
	(112)		(165)	(1)	(13)
170. Nizar	204	-	19.0	57.1	28.1
	(83)		(141)	(21)	(11)
171. Vyara	224	13.2	24.4	59.0	28.5
	(69)	(108)	(118)	(4)	(10)
172. Valod	332	-	33.1	58.9	32.0
	(34)		(60)	(5)	(5)
173. Bardoli	410	24.4	35.9	58.0	26.9
	(18)	(50)	(44)	(9)	(12)
174. Mahuva	294	-	35.8	58.4	33.2
	(38)		(45)	(8)	(2)
175. Palsana	379	-	34.0	59.6	32.7
	(25.5)		(56)	(2.5)	(4)
176. Valsad	555	37.0	49.8	50.9	13.5
	(9)	(21)	(6)	(138.5)	(60.5)
177. Pardi	483	20.5	39.0	53.1	18.2
	(12)	(67)	(30)	(70)	(28)
178. Umbergaon	397	7.6	26.4	51.1	23.3
	(21)	(134)	(106)	(131)	(17)
179. Gandevi	674	36.5	52.3	51.9	13.9
	(6)	(22)	(3)	(109)	(58)
180. Navsari	486	37.6	49.8	53.9	17.3
	(11)	(19)	(7)	(51.5)	(32)
181. Chikhli	373	2.8	37.0	53.9	22.1
	(28)	(143)	(38)	(51.5)	(21)
182. Bansda	243	5.9	21.4	55.9	22.7
	(62)	(139)	(130)	(31)	(20)
183. Dharampur	144	6.0	13.0	57.3	24.3
	(125)	(138)	(163)	(18.5)	(14.5)
184. Dangs	64	-	21.0	56.3	22.8
	(176)		(131)	(27)	(19)
Gujarat State	174	31.1	32.3	52.2	11.0

Note: Figures in brackets indicate ranks which are given on the basis of two decimal figures.

APPENDIX III

TALUKAWISE INDICATORS OF SOCIAL WELL-BEING

Taluka	*No. of primary schools per 10,000 students*	*No. of primary teachers per 10,000 students*	*No. of doctors per 100,000 population*	*No. of beds per 10,000 population*	*No. of post-boxes per 10,000 population*	*Percentage of towns and villages electrified*	*Length of road per 10,000 sq km (in km)*	*No. of cinema halls per 100,000 population*
1	2	3	4	5	6	7	8	9
1. Jamnagar	35	289	16	26	6	100	194	1.9
	(159)	(38)	(6)	(2)	(143)	(14.5)	(157)	(47.5)
2. Lalpur	75	260	7	0.7	4	100	85	1.1
	(48)	(84)	(70.5)	(164.5)	(178)	(14.5)	(182)	(104.5)
3. Jamjodhpur	55	326	5	0.5	8	93	108	0.9
	(96)	(24)	(130)	(170)	(95.5)	(58.5)	(179)	(129)
4. Bhanvad	62	277	4	0.6	8	59	255	1.1
	(81)	(52)	(139.5)	(167)	(88)	(141)	(129)	(112)
5. Kalyanpur	75	238	3	0.5	6	49	173	1.66
	(49)	(133)	(161)	(171)	(149)	(153.5)	(167)	(62)
6. Okhamandal	61	303	9	2	6	33	508	4.28
	(84)	(30)	(34)	(97)	(134)	(169.5)	(42)	(2)

(*Appendix III Contd.*)

1	*2*	*3*	*4*	*5*	*6*	*7*	*8*	*9*
7. Khambata	71	270	5	5	10	100	147	(1.35)
	(54)	(59)	(108)	(44)	(64)	(14.5)	(172)	(90)
8. Jodiya	52	356	6	0.7	5	100	205	1.20
	(103)	(14)	(93)	(163)	(165)	(14.5)	(153)	(102)
9. Dhrol	62	332	9	4	8	100	204	1.72
	(78)	(21)	(39)	(53.5)	(103.5)	(14.5)	(154)	(56.5)
10. Kalavad	52	265	3	0.5	8	86	155	0.79
	(105)	(71)	(163)	(173)	(83)	(82)	(171)	(134)
11. Rajkot	33	228	12	15	2	81	255	2.47
	(164)	(141)	(15)	(6)	(184)	(93)	(130)	(16.5)
12. Kotda Sangani	50	1285	6	0.8	7	83	271	-
	(107)	(1)	(96)	(161)	(110)	(91.5)	(119)	
13. Gondal	30	279	6	4	7	95	360	2.35
	(172)	(49)	(101)	(59)	(111)	(49)	(78)	(20)
14. Jetpur	27	188	5	2	5	84	243	2.30
	(175)	(176)	(117)	(121.5)	(159.5)	(88)	(139)	(25)
15. Dhoraji	27	200	5	3	6	100	335	1.98
	(178)	(169)	(112.5)	(74.5)	(140.5)	(14.5)	(93)	(42)
16. Upleta	13	147	6	4	6	84	254	2.34
	(184)	(183)	(78.5)	(57)	(152)	(88)	(132)	(22)
17. Jamkandorna	44	462	6	1	4	100	132	1.57
	(129)	(5)	(83)	(136)	(172)	(14.5)	(175)	(72)
18. Lodhika Mahal	64	830	11	2	4	79	201	-
	(72)	(3)	(23)	(98)	(177)	(99)	(155)	
19. Paddhari	58	217	6	1	11	76	271	1.53
	(91)	(152)	(89)	(139)	(48)	(109.5)	(117)	(76.5)
20. Morvi	47	77	7	4	6	89	268	1.63
	(117)	(184)	(68)	(62.5)	(132)	(70.5)	(121)	(65.5)

(*Appendix III Contd.*)

1	*2*	*3*	*4*	*5*	*6*	*7*	*8*	*9*
21. Maliya	60	299	10	2	10	98	188	-
	(87)	(32)	(29)	(108)	(60)	(35.5)	(159)	
22. Wankaner	70	386	10	6	11	49	188	1.63
	(58)	(9)	(27)	(41)	(57)	(153.5)	(160)	(65.5)
23. Jasdan	54	189	5	2	6	70	217	0.59
	(98)	(175)	(110)	(103)	(130.5)	(123.5)	(149)	(153)
24. Wadhwan	28	203	18	23	7	98	310	2.90
	(174)	(165)	(4)	(3)	(121.5)	(35.5)	(96)	(9)
25. Limbdi	48	270	15	13	7	65	176	1.68
	(115)	(60)	(9)	(10)	(112)	(135.5)	(169)	(59)
26. Sayla	106	263	7	1	9	79	213	-
	(14)	(77)	(77)	(133.5)	(75)	(99)	(151)	
27. Chotila	103	228	4	0.81	15	40	234	2.03
	(16)	(142)	(146)	(158)	(23)	(162.5)	(142)	(39)
28. Muli	70	261	7	1	6	86	248	-
	(61)	(83)	(59)	(142.5)	(150.5)	(82)	(137)	
29. Halvad	63	260	7	1	11	77	369	1.11
	(74)	(85)	(74.5)	(145)	(56)	(105.5)	(74)	(108.5)
30. Dhrangdhra	41	252	12	9	6	80	277	1.46
	(137.5)	(101)	(16)	(18)	(145)	(95.5)	(109)	(84.5)
31. Dasada	51	265	4	0.58	7	59	285	0.73
	(106)	(72)	(132)	(169)	(113)	(141)	(104)	(139.5)
32. Lakhtar	59	285	7	2	15	58	281	1.78
	(90)	(43)	(64)	(111)	(28)	(143.5)	(106)	(55)
33. Bhavnagar	32	240	8	14	5	73	168	2.46
	(169)	(131)	(43)	(7)	(154)	(117)	(170)	(18.5)
34. Ghogho Mahal	66	269	5	9	6	72	519	-
	(66)	(62)	(126)	(17)	(130.5)	(119.5)	(40)	

(*Appendix III Contd.*)

1	*2*	*3*	*4*	*5*	*6*	*7*	*8*	*9*
35. Talaja	63	179	3	0.82	8	72	387	0.55
	(73)	(178)	(162)	(157)	(97.5)	(119.5)	(69)	(154)
36. Mahuva	52	236	7	6	5	75	264	1.51
	(104)	(138)	(70.5)	(34)	(156.5)	(112.5)	(126)	(80)
37. Kundla	35	252	3	4	8	93	266	0.96
	(160)	(103)	(166)	(61)	(90)	(58.5)	(123)	(121)
38. Gariadhar	40	249	4	2	6	96	266	1.12
	(142)	(110)	(131)	(92)	(146)	(43.5)	(124)	(106.5)
39. Palitana	57	244	5	0.43	7	91	403	2.15
	(94.5)	(124)	(123)	(176)	(118)	(63.5)	(63)	(34.5)
40. Sihor	54	266	13	58	11	89	276	0.75
	(99)	(70)	(12)	(1)	(58)	(70.5)	(110)	(136)
41. Umrala	43	238	6	2	9	86	287	5.52
	(131)	(134)	(106)	(106)	(77.5)	(82)	(102)	(1)
42. Gadhada	47	240	2	0.48	9	96	254	1.62
	(116)	(132)	(170)	(72)	(74)	(43.5)	(133)	(67)
43. Botad	41	238	5	2	6	100	267	1.51
	(141)	(136)	(112.5)	(90)	(144)	(14.5)	(122)	(80)
44. Valabhipur	52	207	11	8	10	100	264	1.58
Mahal	(102)	(158)	(19)	(21)	(70)	(14.5)	(125)	(70)
45. Amreli	27	238	15	11	6	100	307	2.27
	(179)	(135)	(7)	(12)	(137.5)	(14.5)	(98)	(26)
46. Dhari	45	242	6	0.8	4	67	282	1.58
	(124)	(126)	(104.5)	(159.5)	(175)	(130)	(105)	(70)
47. Khambha	49	262	11	6	7	69	339	2.15
Mahal	(113)	(79)	(22)	(40)	(117)	(126)	(91)	(34.5)
48. Rajula	66	254	6	6	4	63	384	1.51
	(67)	(98)	(90.5)	(43)	(173)	(138)	(72)	(80)

(*Appendix III Contd.*)

1	*2*	*3*	*4*	*5*	*6*	*7*	*8*	*9*
49. Jafrabad	71	162	5	2	7	47	354	3.38
	(55)	(181)	(121)	(110)	(119.5)	(156)	(81)	(4)
50. Kodinar	38	193	4	0.9	8	100	391	0.72
	(153)	(174)	(155.5)	(154.5)	(105)	(14.5)	(67)	(142)
51. Kunkavav Vadia	29	218	5	2	9	84	250	1.96
	(173)	(151)	(129)	(117.5)	(79)	(88)	(135)	(43)
52. Babra	36	221	4	0.7	12	66	272	2.18
	(158)	(149)	(138)	(167)	(41.5)	(132.5)	(115)	(30.5)
53. Lathi	33	248	8	5	13	80	345	1.99
	(162)	(116)	(48)	(45)	(38)	(95.5)	(87)	(41)
54. Lilia Mahal	39	248	9	1	15	84	301	1.86
	(152)	(113)	(32)	(126)	(26)	(88)	(100)	(50)
55. Junagadh	26	251	14	16	6	87	231	3.29
	(181)	(106)	(11)	(4)	(153)	(77.5)	(144)	(5)
56. Mandarda	39	240	9	3	11	89	305	1.72
	(149)	(130)	(40)	(70)	(51)	(70.5)	(99)	(56.5)
57. Talala	32	204	8	2	10	52	145	2.18
	(166)	(164)	(52)	(125)	(69)	(151)	(173)	(30.5)
58. Patan Veraval	39	202	3	3	5	57	199	1.54
	(148)	(167)	(165)	(66)	(164)	(145.5)	(156)	(74.5)
59. Malia	40	206	6	1	6	95	261	1.80
	(143)	(161)	(82)	(129)	(134)	(49)	(127)	(53.5)
60. Mangrol	38	206	3	1	5	92	187	0.70
	(156)	(159)	(157)	(137.5)	(159.5)	(61.5)	(161)	(143.5)
61. Keshod	27	244	5	2	14	96	207	2.19
	(176)	(122)	(118)	(121.5)	(35)	(43.5)	(152)	(29)
62. Porbandar	33	226	8	8	5	90	130	2.21
	(165)	(145)	(41)	(20)	(163)	(66)	(176)	(28)

(Appendix III Contd.)

1	*2*	*3*	*4*	*5*	*6*	*7*	*8*	*9*
63. Ranavav	42	237	7	2	8	55	224	—
	(136)	(137)	(74.5)	(90)	(93)	(148.5)	(148)	(—)
64. Kutiyana	45	242	9	2	7	94	238	1.25
	(125)	(125)	(38)	(90)	(115)	(54)	(140)	(97.5)
65. Manavadar	31	299	4	1	57	96	237	1.60
	(171)	(31)	(148)	(150)	(2)	(43.5)	(141)	(68)
66. Vanthali	32	258	7	1	8	94	216	1.04
	(167)	(88)	(61)	(137.5)	(100)	(54)	(150)	(115)
67. Bhesan Mahal	33	210	8	1	9	80	520	1.50
	(163)	(155)	(55)	(141)	(72.5)	(95.5)	(39)	(82)
68. Visavadar	39	208	4	1	5	78	175	0.85
	(146)	(157)	(141)	(132)	(158)	(102)	(165).	(131)
69. Una	57	203	2	1	7	65	95	1.34
	(94.5)	(166)	(174)	(146.5)	(108.5)	(135.5)	(180)	(91)
70. Bhuj	99	344	6	2	8	72	126	2.31
	(20)	(18)	(103)	(13)	(94)	(119.5)	(177)	(24)
71. Mundra	85	254	17	5	9	86	368	1.58
	(38)	(99)	(5)	(49)	(71)	(82)	(76)	(70)
72. Mandvi	60	206	9	1	8	97	368	1.54
	(88)	(162)	(33)	(133.5)	(86)	(39)	(77)	(74.5)
73. Abdasa	125	251	27	6	19	63	324	—
	(6)	(104)	(1)	(32)	(13)	(138)	(95)	(—)
74. Lakhpat	137	278	24	2	29	11	169	—
	(3)	(50)	(2)	(100)	(5)	(183.5)	(169)	(—)
75. Nakhatrana	61	272	8	2	12	63	272	0.94
	(83)	(56)	(53)	(117.5)	(44)	(138)	(116)	(125)
76. Rapar	80	225	9	0.4	8	57	123	0.74
	(43)	(147)	(37)	(175)	(87)	(145.5)	(178)	(137.5)

(*Appendix III Contd.*)

1	2	3	4	5	6	7	8	9
77. Bhachav	82	255	11	2	8	66	84	0.99
	(40)	(97)	(20)	(105)	(97.5)	(132.5)	(183)	(118)
78. Anjar	63	241	5	2	5	77	226	2.61
	(76)	(128)	(127)	(85)	(155)	(105.5)	(147)	(14)
79. Palanpur	50	250	9	7	6	89	279	1.30
	(111)	(107)	(31)	(26)	(147)	(70.5)	(108)	(93.5)
80. Vadgam	47	270	6	—	6	96	391	0.73
	(118)	(61)	(97)	(–)	(140.5)	(43.5)	(66)	(140)
81. Deesa	62	208	2	1	7	74	234	1.15
	(82)	(156)	(171)	(128)	(124)	(115)	(143)	(103)
82. Kankrej	73	224	7	—	7	91	351	0.68
	(52)	(148)	(72)	(–)	(129)	(63.5)	(83)	(148)
83. Radhanpur	70	227	6	2	13	59	487	2.46
	(59)	(144)	(87)	(93.5)	(36)	(141)	(44)	(18.5)
84. Santalpur	95	196	10	—	16	28	273	—
	(24.5)	(171)	(28)	(–)	(21)	(172)	(114)	(–)
85. Deodar	96	197	4	—	10	83	274	1.38
	(23)	(170)	(144)	(–)	(59)	(91.5)	(113)	(89)
86. Vav	92	183	5	—	8	35	349	—
	(29)	(177)	(124)	(–)	(103.5)	(166.5)	(84)	(–)
87. Tharad	101	174	5	3	8	44	339	0.68
	(17)	(180)	(125)	(77)	(106)	(160.5)	(90)	(148)
88. Dhanera	95	212	4	—	12	45	228	—
	(27)	(153)	(142)	(–)	(43)	(158.5)	(146)	(–)
89. Danta	123	258	8	2	21	46	352	0.98
	(7)	(90)	(49)	(81)	(11)	(157)	(82)	(119)
90. Himatnagar	62	298	10	10	11	88	383	1.06
	(77)	(33)	(30)	(14)	(55)	(75)	(73)	(113.5)

(*Appendix III Contd.*)

1	*2*	*3*	*4*	*5*	*6*	*7*	*8*	*9*
91. Prantij	65	193	6	1	9	94	517	0.92
	(69.5)	(173)	(78.5)	(67)	(72.5)	(54)	(41)	(127.5)
92. Idar	59	267	7	6	10	97	331	1.24
	(89)	(66)	(56.5)	(36)	(66)	(39)	(94)	(99.5)
93. Khedbrahma	122	569	8	3	14	54	229	0.74
	(8)	(4)	(45)	(71)	(32)	(150)	(145)	(137.5)
94. Vijaynagar	71	247	10	5	19	89	476	—
Mahal	(56)	(117)	(24)	(47)	(14)	(70.5)	(48)	(—)
95. Bhiloda	69	274	7	12	15	71	583	0.68
	(62)	(55)	(69)	(11)	(27)	(122)	(29)	(148)
96. Meghraj	99	218	8	2	18	68	748	—
	(21)	(150)	(51)	(84)	(17.5)	(128.5)	(16)	(—)
97. Modasa	80	303	10	10	12	77	721	2.08
	(41)	(29)	(25)	(15)	(41.5)	(105.5)	(18)	(37)
98. Malpur	100	1174	6	1	19	86	780	—
	(19)	(2)	(86)	(152)	(15)	(82)	(14)	(—)
99. Bayad	95	384	7	4	14	77	588	1.80
	(24.5)	(10)	(76)	(60)	(30)	(105.5)	(27)	(53.5)
100. Mehsana	44	262	11	9	7	100	186	0.65
	(127)	(81)	(21)	(19)	(119.5)	(14.5)	(162)	(150)
101. Kadi	46	258	5	3	4	94	193	0.92
	(120)	(89)	(120)	(78)	(168)	(54)	(158)	(127.5)
102. Chanasma	40	252	4	2	6	97	182	0.96
	(145)	(102)	(151)	(115)	(139)	(39)	(163)	(121)
103. Sami	57	206	4	1	10	66	90	—
	(93)	(160)	(149.5)	(151)	(67)	(132.5)	(181)	(—)
104. Harij	65	250	7	2	9	95	275	1.45
	(69.5)	(108)	(62)	(113.5)	(76)	(49)	(111)	(86.5)

(*Appendix III Contd.*)

1	*2*	*3*	*4*	*5*	*6*	*7*	*8*	*9*
105. Patan	43	161	5	4	4	99	139	1.65
	(132)	(182)	(114.5)	(53.5)	(167)	(31.5)	(174)	(63.5)
106. Sidhpur	27	251	5	6	4	100	173	1.39
	(177)	(105)	(116)	(33)	(170)	(14.5)	(166)	(88)
107. Kheralu	53	267	4	1	7	89	172	1.22
	(101)	(67)	(154)	(149)	(126)	(70.5)	(168)	(101)
108. Visnagar	39	256	6	6	6	100	274	1.01
	(151)	(96)	(104.5)	(29)	(142)	(14.5)	(112)	(116.5)
109. Vijapur	38	248	6	4	5	99	249	1.65
	(155)	(115)	(90.5)	(62.5)	(162)	(31.5)	(136)	(63.5)
110. Kalol	34	211	6	6	8	96	341	1.82
	(161)	(154)	(80.5)	(42)	(89)	(43.5)	(88)	(51.5)
111. Gandhinagar	31	263	15	6	7	69	455	0.69
	(170)	(76)	(8)	(35)	(107)	(126)	(54)	(145.5)
112. Ahmedabad	23	232	23	14	2	100	4589	2.17
City	(182)	(140)	(3)	(8)	(183)	(14.5)	(1)	(32)
113. Daskroi	37	246	2	0.9	4	86	incl. in	0.36
	(157)	(119)	(178)	(154.5)	(176)	(82)	A'bad	(159.5)
114. Dholka	41	276	6	4	7	100	348	1.53
	(139)	(54)	(88)	(58)	(127.5)	(14.5)	(85)	(76.5)
115. Dhandhuka	45	289	7	4	7	72	253	1.87
	(123)	(37)	(66)	(52)	(114)	(119.5)	(134)	(49)
116. Sanand	47	264	4	2	4	90	467	0.73
	(119)	(74)	(134)	(102)	(174)	(86)	(50)	(140)
117. Viramgam	50	263	3	2	6	78	340	1.11
	(108)	(75)	(168)	(101)	(137.5)	(102)	(89)	(108.5)
118. Dehgam	54	263	4	0.8	3	98	569	1.09
	(100)	(78)	(136)	(162)	(180)	(35.5)	(35)	(111)

(*Appendix III Contd.*)

1	*2*	*3*	*4*	*5*	*6*	*7*	*8*	*9*
119. Mehemdabad	43 (130)	348 (17)	7 (65)	7 (24)	7 (123)	100 (14.5)	651 (23)	0.94 (125)
120. Nadiad	91 (31)	282 (47)	4 (155.5)	7 (27)	5 (161)	100 (14.5)	888 (7)	1.91 (46)
121. Anand	43 (133)	427 (7)	1 (179)	3 (68.5)	4 (166)	100 (14.5)	962 (4)	2.34 (22)
122. Petlad	39 (147)	438 (6)	3 (167)	5 (46)	4 (170)	100 (14.5)	932 (5)	1.31 (92)
123. Borsad	41 (140)	351 (16)	0.8 (183)	2 (86)	3 (182)	100 (14.5)	823 (10)	0.81 (133)
124. Khambat	40 (144)	336 (20)	6 (96)	7 (28)	6 (136)	100 (14.5)	580 (31)	1.52 (78)
125. Matar	46 (121)	330 (23)	— (—)	— (—)	6 (148)	100 (14.5)	6[illegible] (20)	1.24 (99.5)
126. Kapadvanj	62 (80)	296 (34)	0.9 (182)	2 (123)	9 (81)	100 (14.5)	534 (37)	0.94 (125)
127. Balasinor	80 (42)	353 (15)	2 (169)	5 (50)	10 (68)	100 (14.5)	574 (32)	0.62 (151)
128. Thasra	68 (63)	342 (19)	1 (180)	2 (107)	7 (121.5)	100 (14.5)	444 (58)	1.28 (95)
129. Godhra	77 (46)	393 (8)	6 (102)	6 (30)	8 (84)	35 (166.5)	446 (56)	0.88 (130)
130. Kalol	66 (65)	283 (44)	3 (158.5)	0.8 (156)	11 (49)	76 (109.5)	453 (55)	0.69 (145.5)
131. Halol	105 (15)	201 (168)	2 (172)	0.5 (174)	15 (24)	75 (112.5)	399 (65)	0.76 (135)
132. Shehera	67 (64)	265 (73)	2 (176)	0.4 (177)	10 (63)	18 (181)	552 (36)	— (—)

(*Appendix III Contd.*)

1	*2*	*3*	*4*	*5*	*6*	*7*	*8*	*9*
133. Lunawada	78	286	2	2	8	36	618	1.30
	(45)	(42)	(175)	(96)	(102)	(165)	(25)	(93.5)
134. Santrampur	91	291	4	1	8	30	585	0.31
	(32)	(36)	(139.5)	(127)	(101)	(171)	(28)	(161)
135. Jhalod	76	194	5	2	9	26	683	0.50
	(47)	(172)	(107)	(119)	(80)	(173)	(21)	(156)
136. Dohad	63	331	4	2	8	56	571	1.01
	(75)	(22)	(147)	(83)	(85)	(147)	(34)	(116.5)
137. Limkheda	112	244	6	2	7	24	527	—
	(10)	(121)	(100)	(88)	(108.5)	(175.5)	(38)	(—)
138. Devgadbaria	94	258	4	2	5	21	406	0.36
	(28)	(91)	(136)	(113.5)	(156.5)	(178.5)	(60)	(159.5)
139. Jambughoda	145	261	12	2	60	95	445	—
	(1)	(80)	(14)	(82)	(1)	(49)	(57)	(—)
140. Vadodara	26	257	10	15	3	98	399	1.92
	(180)	(94)	(26)	(5)	(179)	(35.5)	(64)	(45)
141. Karjan	50	175	5	1	7	99	387	1.46
	(110)	(179)	(119)	(142.5)	(127.5)	(31.5)	(70)	(84.5)
142. Padra	42	259	3	1	6	93	456	1.49
	(135)	(87)	(158.5)	(135)	(150.5)	(58.5)	(53)	(83)
143. Savli	65	249	4	2	11	79	337	0.52
	(68)	(111)	(145)	(112)	(52)	(99)	(92)	(155)
144. Vaghodia	85	249	6	1	17	80	404	0.95
	(37)	(109)	(98)	(130)	(19)	(95.5)	(62)	(123)
145. Dabhoi	61	266	7	4	12	84	299	1.82
	(85)	(68)	(59)	(65)	(47)	(88)	(101)	(51.5)
146. Sankheda	91	261	5	1	11	66	285	2.01
	(30)	(82)	(109)	(140)	(54)	(132.5)	(103)	(40)

(*Appendix III Contd.*)

1	2	3	4	5	6	7	8	9
147. Jetpur Pavi	100	253	4	1	11	55	258	0.60
	(18)	(100)	(143)	(146)	(50)	(148.5)	(128)	(152)
148. Chhota	132	308	9	6	6	33	254	0.42
Udaipur	(4)	(27)	(35)	(39)	(134)	(169.5)	(131)	(158)
149. Nasvadi	112	249	7	3	21	20	271	1.12
	(11)	(112)	(73)	(72)	(12)	(180)	(118)	(106.5)
150. Tilakwada	83	233	1	7	21	70	487	—
Mahal	(39)	(139)	(181)	(23)	(10)	(123.5)	(45)	(—)
151. Sinor	42	248	12	2	3	100	390	2.89
	(134)	(114)	(13)	(87)	(181)	(14.5)	(68)	(10)
152. Bharuch	38	247	8	6	12	74	457	2.26
	(154)	(118)	(54)	(38)	(45)	(115)	(52)	(27)
153. Ankleshwar	45	306	3	5	13	74	574	1.55
	(126)	(28)	(164)	(51)	(40)	(115)	(33)	(73)
154. Hansot	70	376	8	3	10	58	484	—
	(60)	(11)	(47.5)	(73)	(65)	(143.5)	(47)	(—)
155. Vagra	62	375	5	1	14	40	268	—
	(79)	(12)	(128)	(131)	(31)	(162.5)	(120)	(—)
156. Jambusar	45	226	5	3	9	44	385	1.26
	(122)	(146)	(122)	(67)	(77.5)	(160.5)	(71)	(96)
157. Amod	50	288	4	2	10	92	359	1.25
	(112)	(40)	(152)	(104)	(62)	(61.5)	(79)	(97.5)
158. Jhagadia	91	271	6	2	15	48	369	2.34
	(33)	(57)	(85)	(120)	(25)	(155)	(75)	(22)
159. Nandod	112	287	4	6	13	45	310	1.67
	(9)	(41)	(150)	(31)	(37)	(158.5)	(97)	(60.5)
160. Dediapada	127	246	2	0.7	14	24	280	1.13
	(5)	(120)	(173)	(164.5)	(29)	(175.5)	(107)	(104.5)

(*Appendix III Contd.*)

1	*2*	*3*	*4*	*5*	*6*	*7*	*8*	*9*
161. Sagbara	95	227	3	1	18	24	355	1.71
	(26)	(143)	(160)	(148)	(16)	(175.5)	(80)	(58)
162. Valia	90	317	6	4	10	68	422	1.0[illegible]
	(34)	(26)	(80.5)	(55)	(61)	(128.5)	(59)	(113.5)
163. Chorasi	22	204	8	10	4	87	495	2.67
	(183)	(163)	(42)	(16)	(170)	(77.5)	(43)	(13)
164. Olpad	50	269	6	2	13	76	663	2.98
	(109)	(63)	(94)	(80)	(39)	(109.5)	(22)	(8)
165. Kamraj	61	244	7	3	16	100	753	2.77
	(86)	(123)	(58)	(74.5)	(22)	(14.5)	(17)	(11)
166. Mangrol	79	266	8	2	12	51	593	2.47
	(44)	(69)	(46)	(93.5)	(46)	(152)	(26)	(16.5)
167. Mandvi	89	256	4	4	17	37	471	1.45
	(35)	(95)	(136)	(64)	(20)	(164)	(49)	(86.5)
168. Songadh	112	277	2	2	22	24	247	2.05
	(12)	(53)	(177)	(95)	(9)	(175.5)	(138)	(38)
169. Uchchhal	96	278	11	2	25	12	461	1.90
	(22)	(51)	(18)	(116)	(7)	(182)	(51)	(47.5)
170. Nizar	87	258	7	1	26	69	346	2.48
	(36)	(92)	(56.5)	(144)	(6)	(126)	(86)	(15)
171. Vyara	75	268	7	2	14	34	484	1.10
	(50)	(64)	(63)	(124)	(34)	(168)	(46)	(110)
172. Valod	71	257	6	0.9	34	78	696	3.02
	(57)	(93)	(92)	(153)	(3)	(102)	(19)	(7)
173. Bardoli	44	242	14	7	23	88	774	3.22
	(128)	(127)	(10)	(22)	(8)	(75)	(15)	(6)
174. Mahuva	74	283	8	2	18	88	870	0.96
	(51)	(45)	(50)	(79)	(17.5)	(75)	(8)	(121)

(*Appendix III Contd.*)

1	*2*	*3*	*4*	*5*	*6*	*7*	*8*	*9*
175. Palsana	57 (92)	260 (86)	5 (114.5)	0.8 (159.5)	14 (33)	100 (14.5)	1001 (3)	3.94 (3)
176. Valsad	41 (137.5)	282 (46)	9 (36)	7 (25)	8 (99)	86 (82)	925 (6)	2.16 (33)
177. Pardi	49 (114)	271 (58)	4 (133)	3 (76)	8 (92)	99 (31.5)	798 (12)	1.95 (44)
178. Umbergaon	65 (71)	267 (65)	6 (84)	3 (68.5)	8 (95.5)	94 (54)	831 (9)	0.70 (143.5)
179. Gandevi	39 (150)	360 (13)	4 (153)	5 (48)	7 (116)	95 (49)	787 (13)	2.10 (36)
180. Navsari	32 (168)	241 (129)	5 (111)	4 (56)	8 (91)	90 (66)	640 (24)	1.67 (60.5)
181. Chikhli	55 (97)	296 (35)	7 (67)	2 (99)	8 (82)	93 (58.5)	1022 (2)	0.47 (157)
182. Bansda	72 (53)	289 (39)	8 (44)	6 (37)	7 (125)	76 (109.5)	801 (11)	2.74 (12)
183. Dharampur	111 (13)	281 (48)	6 (95)	2 (109)	11 (53)	21 (178.5)	406 (61)	0.84 (132)
184. Dangs	141 (2)	322 (25)	11 (17)	13 (9)	31 (4)	11 (183.5)	581 (30)	— —

Note: Figures in brackets indicate ranks which are given on the basis of the data upto two decimals.

APPENDIX IV

RANK ORDER OF TALUKAS ON THE BASIS OF THE LEVEL OF SOCIAL WELL-BEING

(Micro-regions)

	Micro-region/ Taluka	*Total weight-age*	*Levels of Social well-being*	*Weight-age for social indi-cators*	*Weight-age of demo-graphic indi-cators*
	1	*2*	*3*	*4*	*5*
I.	**Kathiawar (North) Kachchh**				
1.	Rajkot	68.3	VH	48.6	19.6
2.	Jamnagar	60.2	VH	42.2	17.9
3.	Wadhwan	51.4	H	32.9	18.5
4.	Morvi	47.7	H	32.4	15.3
5.	Anjar	47.0	H	31.7	15.4
6.	Wankaner	40.9	M	26.7	14.2
7.	Viramgam	38.9	M	23.4	15.5
8.	Dhrangdhra	38.9	M	23.5	15.4
9.	Limbdi	38.9	M	23.1	15.8
10.	Okhamandal	37.8	M	23.1	14.6
11.	Paddhari	36.5	M	21.9	14.5
12.	Jamjodhpur	34.9	M	19.5	15.5
13.	Bhuj	33.8	M	19.3	14.6
14.	Mandvi	32.6	M	17.2	15.4
15.	Khambata	32.5	M	19.7	12.8
16.	Dhrol	32.4	M	17.8	14.6
17.	Bhanvad	30.4	L	15.2	15.2
18.	Maliya	30.0	L	15.4	14.6
19.	Kalavad	29.8	L	15.8	14.0
20.	Jodiya	29.7	L	14.9	14.8
21.	Lodhika Mahal	29.2	L	15.9	13.3
22.	Halwad	28.2	L	13.3	14.9
23.	Muli	28.0	L	12.5	15.5
24.	Lalpur	27.0	L	14.8	12.2
25.	Lakhtar	26.9	L	10.1	16.8
26.	Chotila	26.4	L	12.9	13.4
27.	Dasada	26.1	L	12.2	13.9

(Appendix IV Contd.)

1		2	3	4	5
28.	Nakhatrana	26.0	L	10.7	15.3
29.	Mundra	25.9	L	11.2	14.7
30.	Kalyanpur	25.5	L	11.9	13.6
31.	Sayla	24.1	L	9.9	14.2
32.	Bachau	23.1	L	10.1	12.9
33.	Rapar	22.6	L	10.5	12.1
34.	Abdasa	21.1	L	6.4	14.7
35.	Lakhpat	18.5	L	5.5	13.0
II.	**Kathiawar Core**				
1.	Junagadh	58.1	VH	41.0	17.1
2.	Upleta	47.9	H	30.9	17.0
3.	Jetpur	47.3	H	32.5	14.8
4.	Dhoraji	46.3	H	30.1	16.2
5.	Amreli	45.3	H	30.1	16.2
6.	Gondal	43.0	M	28.5	14.5
7.	Sihor	40.9	M	26.0	14.9
8.	Visavadar	37.9	M	22.9	15.0
9.	Botad	36.3	M	20.8	15.5
10.	Kundla	34.2	M	19.8	14.4
11.	Jasdan	32.4	M	19.6	12.8
12.	Kunkavav	29.7	L	13.3	16.4
13.	Gariadhar	29.6	L	14.1	15.5
14.	Jamkondra	29.4	L	16.0	13.4
15.	Lathi	28.9	L	12.3	16.6
16.	Liliamahal	28.3	L	11.7	16.6
17.	V.B. Mahal	28.0	L	13.8	14.2
18.	K.S. Mahal	27.7	L	14.5	13.2
19.	Palitana	27.7	L	14.3	13.4
20.	Gadharda	27.5	L	13.9	13.6
21.	Bhesan Mahal	26.9	L	12.7	14.2
22.	Dhari	26.1	L	11.8	14.3
23.	Babra	25.6	L	11.3	14.3
24.	Umrala	24.3	L	10.3	14.0
25.	Khamba Mahal	23.4	L	11.4	12.0
III.	**Kathiawar Coastal**				
1.	Bhavnagar	47.5	VH	31.5	16.0
2.	Talala	41.5	H	20.4	21.1
3.	Patan Veraval	40.0	H	25.5	15.5
4.	Keshod	38.7	H	23.5	15.2
5.	Porbandar	38.6	H	23.8	14.8
6.	Mahuva	35.2	M	21.5	13.7
7.	Manavadar	34.7	M	18.4	16.3
8.	Vanthali	33.5	M	18.3	15.2
9.	Mangrol	32.8	M	18.3	14.5
10.	Mendarda	32.7	M	13.7	19.0
11.	Una	32.4	M	19.8	12.6
12.	Maliya	31.4	M	17.0	14.4

(Appendix IV Contd.)

1	2	3	4	5
13. Kutiyana	30.8	M	14.6	16.2
14. Rajula	28.9	M	15.5	13.4
15. Kodinar	28.1	M	14.7	13.4
16. Talaja	27.3	M	13.7	13.6
17. Jafrabad	24.5	L	11.8	12.7
18. Ranavav	24.1	L	9.5	14.6
19. G. Mahal	23.1	L	9.3	13.8
IV. North Gujarat Region				
1. Visnagar	50.6	VH	32.2	14.4
2. Sidhpur	50.3	VH	32.4	17.9
3. Mehsana	48.9	VH	31.1	17.8
4. Kalol	47.6	VH	30.6	17.0
5. Palanpur	46.6	H	32.2	14.4
6. Vijapur	43.0	H	25.4	17.6
7. Deesa	41.8	H	24.8	17.0
8. Kadi	41.4	H	24.2	17.2
9. Patan	40.6	M	20.1	18.5
10. Kheralu	36.1	M	20.1	16.0
11. Tharad	32.2	M	20.6	11.6
12. Vadgam	29.8	M	15.0	14.8
13. Chansma	29.5	M	12.9	16.6
14. Radhanpur	28.6	M	14.7	13.8
15. Harij	27.6	M	13.1	14.5
16. Danta	26.9	M	15.9	11.0
17. Kankrej	25.5	M	13.4	12.1
18. Deedar	25.4	M	12.4	13.0
19. Santalpur	21.6	L	8.9	12.7
20. Dhanera	21.4	L	9.4	11.9
21. Sami	20.6	L	8.6	12.0
22. Vav	20.0	L	9.6	10.4
V. Central Gujarat Region				
1. Ahmedabad	82.4	VH	54.2	28.2
2. Nadiad	39.8	H	25.4	14.4
3. Anand	39.6	H	25.5	14.1
4. Modasa	39.1	H	24.2	14.8
5. Himatnagar	38.4	H	23.7	14.7
6. Dholka	36.4	M	19.9	16.6
7. Dehgam	36.3	M	21.5	14.7
8. Gandhinagar	36.2	M	22.2	14.0
9. Mehmadabad	35.4	M	21.5	13.9
10. Petlad	35.1	M	21.1	14.0
11. Kapadvanj	34.8	M	21.2	13.7
12. Daskroi	33.4	M	19.7	13.7
13. Idar	33.4	M	18.9	14.5
14. Balasinor	32.9	M	19.1	13.7
15. Prantij	32.0	M	17.8	14.2
16. Sanand	31.6	M	16.2	15.4

(Appendix IV Contd.)

1		*2*	*3*	*4*	*5*
17.	Thasra	30.0	M	17.4	12.6
18.	Matar	29.1	M	16.3	12.8
19.	Bhiloda	27.2	L	12.5	14.7
20.	Khedbrahma	25.4	L	12.2	13.2
21.	Bayad	24.8	L	12.0	12.8
22.	V. Mahal	24.5	L	10.4	14.1
23.	Malpur	23.2	L	10.3	12.8
24.	Meghraj	22.2	L	9.8	12.4
VI.	**South Gujarat Region**				
1.	Vadodra	85.8	VH	64.9	20.9
2.	Chorasi	63.6	VH	41.9	21.7
3.	Khambat	49.8	H	33.5	16.3
4.	Valsad	47.2	H	29.0	18.2
5.	Chikli	46.9	H	31.2	15.6
6.	Gandevi	44.3	H	26.8	17.5
7.	Navsari	42.4	H	27.0	15.4
8.	Borsad	39.7	H	24.1	15.6
9.	Bardoli	38.3	H	24.7	13.6
10.	Padra	38.1	H	24.9	13.2
11.	Pardi	37.8	H	22.2	15.6
12.	Bharuch	36.2	H	20.8	15.4
13.	Dabhoi	34.0	M	20.0	14.0
14.	Bansda	33.5	M	17.6	15.9
15.	Jambusar	33.2	M	18.1	15.1
16.	Halol	32.6	M	19.1	13.5
17.	Karjan	32.5	M	19.9	12.6
18.	Vyara	32.0	M	17.7	14.3
19.	Dhanduka	31.6	M	17.5	14.1
20.	Jambughoda	30.6	M	13.9	16.7
21.	Ankleshwar	30.5	M	16.5	14.0
22.	Sankheda	30.5	M	16.8	13.7
23.	Umbergaon	29.3	M	13.7	15.6
24.	Kamrej	29.2	M	16.6	12.6
25.	Savli	28.6	M	15.9	12.7
26.	Sinor	28.6	M	15.7	12.9
27.	Olpad	28.1	M	13.7	14.4
28.	Tilakwad Mahal	27.0	L	11.8	15.2
29.	Vaghodia	26.2	L	13.4	12.8
30.	Mandvi	26.2	L	12.3	13.9
31.	Valod	26.0	L	12.1	13.9
32.	Dharmpur	25.8	L	12.5	13.3
33.	Amod	25.7	L	12.6	13.1
34.	Mahuva	25.1	L	12.4	12.8
35.	Hansot	24.1	L	11.3	12.8
36.	Palasara	24.0	L	8.7	15.3
37.	Mangrol	23.9	L	11.3	12.6
38.	Vagra	21.2	L	7.9	13.3

(Appendix IV Contd.)

1		2	3	4	5
VII.	**Eastern Hilly Region**				
1.	Chota Udaipur	53.8	VH	39.8	13.9
2.	Devgadh Baria	50.6	VH	35.5	15.1
3.	Nandod	49.0	VH	32.1	16.9
4.	Dohad	48.0	VH	30.7	17.3
5.	Godhra	45.9	VH	28.3	17.6
6.	Lunawada	37.8	H	22.1	15.7
7.	Kalol	35.3	H	18.9	16.4
8.	Santarampur	33.4	H	19.3	14.1
9.	Limkheda	31.3	M	19.1	12.2
10.	Songadh	31.0	M	14.7	16.3
11.	Shehera	30.6	M	17.9	12.7
12.	Valia	30.1	M	15.4	14.7
13.	Jhalod	30.0	M	16.7	13.3
14.	Jetpur Pavi	29.8	M	16.8	13.0
15.	Jhaghodia	27.7	M	12.6	15.1
16.	Naswadi	26.8	M	14.0	12.8
17.	Sagbara	26.2	M	11.6	14.6
18.	Nizar	26.2	M	11.4	14.8
19.	Dangs	24.6	L	11.1	13.5
20.	Dediapada	24.3	L	10.4	13.9
21.	Uchchchal	23.6	L	9.7	13.9

VH= Very High.
H = High.
M = Moderate.
L = Low levels of well-being.

APPENDIX V

DEMOGRAPHIC, SOCIO-ECONOMIC AND COMPOSITE INDEX OF WELL-BEING

(Talukas arranged in the decreasing order of composite well-being)

	Taluka	*Composite index of well-being*		*Index of well-being based on demo-graphic indicators*	*Index of well-being based on socio-economic indicators*
	1	2		3	4
1.	Bardoli	41.02	VH	14.46	26.56
2.	Mahuva	42.97	VH	12.64	30.33
3.	Palsana	46.30	VH	11.96	34.34
4.	Valod	49.86	VH	14.13	35.73
5.	Kamrej	51.82	VH	24.24	27.58
6.	Valsad	57.10	VH	25.54	31.56
7.	Ankleshwar	58.62	VH	15.82	42.80
8.	Pardi	60.58	H	22.50	38.08
9.	Gandevi	60.79	H	21.52	39.27
10.	Navsari	61.02	H	13.09	47.93
11.	Bharuch	61.72	H	20.76	40.96
12.	Dabhoi	61.99	H	23.64	38.35
13.	Jamnagar	63.77	H	25.27	38.50
14.	Chorasi	65.51	H	17.45	48.06
15.	Olpad	65.79	H	30.16	35.63
16.	Jhagadia	65.87	H	27.01	38.86
17.	Sinor	67.71	H	25.49	42.22
18.	Hansot	68.40	H	33.70	34.70
19.	Dangs	68.96	H	47.96	21.00
20.	Valia	69.66	H	31.79	37.87
21.	Bansda	70.77	H	41.52	29.25
22.	Chikhli	71.18	H	30.59	40.59
23.	Nadiad	72.35	H	39.13	33.22
24.	Bayad	72.73	H	46.47	26.26
25.	Modasa	72.83	H	51.63	21.20
26.	Nizar	72.93	H	34.78	38.15
27.	Uchchhal	73.13	H	39.54	33.59
28.	Vadodara	74.44	H	31.74	42.70

(*Appendix V Contd.*)

	1	*2*		*3*	*4*
29.	Anand	74.64	H	34.29	40.35
30.	Himatnagar	75.11	H	43.15	31.96
31.	Ahmedabad City	76.11	H	37.83	38.28
32.	Jambughoda	76.13	H	54.08	22.05
33.	Mehmedabad	78.26	H	42.83	35.43
34.	Vyara	78.36	H	33.59	44.77
35.	Sankheda	78.87	H	31.96	46.91
36.	Khambat	79.14	H	41.74	37.40
37.	Petlad	79.21	H	35.22	43.99
38.	Amod	79.35	H	31.25	48.10
39.	Lathi	79.40	H	36.63	42.97
40.	Nandod	80.76	MH	41.09	39.67
41.	Wadhwan	81.52	MH	40.22	41.30
42.	Vallabhipur Mahal	81.76	MH	42.39	39.37
43.	Mangrol	82.40	MH	48.91	33.49
44.	Amreli	82.51	MH	41.14	41.37
45.	Mandvi	82.54	MH	38.42	44.12
46.	Dhoraji	82.83	MH	26.85	55.98
47.	Songadh	83.53	MH	40.22	43.31
48.	Wankaner	83.68	MH	44.89	38.79
49.	Dholka	83.98	MH	40.33	43.65
50.	Umbergaon	84.56	MH	44.45	40.11
51.	Gondal	85.42	MH	42.01	43.41
52.	Upleta	86.88	MH	25.98	60.90
53.	Bhuj	86.88	MH	48.26	38.62
54.	Malpur	87.32	MH	58.59	28.73
55.	Vaghodia	87.43	MH	41.68	45.75
56.	Abdasa	87.54	MH	57.34	30.20
57.	Junagadh	87.60	MH	41.30	46.30
58.	Visnagar	87.76	MH	35.76	52.00
59.	Karjan	88.05	MH	28.35	58.70
60.	Matar	88.45	MH	49.86	38.59
61.	Sihor	89.49	MH	51.68	37.81
62.	Rajkot	89.56	MH	38.64	50.92
63.	Mehsana	90.95	MH	43.80	47.15
64.	Idar	91.82	MH	54.73	37.09
65.	Umrala	91.92	MH	41.68	50.24
66.	Godhra	92.07	MH	49.78	42.29
67.	Kalol	92.98	MH	44.78	48.20
68.	Thasra	93.29	MH	48.59	44.70
69.	Dhangadhra	93.32	MH	45.33	47.99
70.	Savli	93.62	MH	36.96	56.66
71.	Lakhpat	93.73	MH	53.94	39.79
72.	Tilakwada Mahal	93.86	MH	50.34	43.52
73.	Ghogho Mahal	94.10	MH	50.54	43.56
74.	Halol	94.29	MH	35.49	58.80
75.	Padra	94.33	MH	35.87	58.46

(*Appendix V Contd.*)

	1	2		3	4
76.	Gandhinagar	94.34	MH	45.32	49.02
77.	Khambha Mahal	94.40	MH	52.11	42.29
78.	Borsad	94.77	MH	42.83	51.94
79.	Vagra	94.85	MH	43.34	51.51
80.	Mundra	94.92	MH	61.63	33.29
81.	Dhrol	95.06	MH	59.73	35.33
82.	Lodhika Mahal	95.28	MH	46.60	48.68
83.	Prantij	95.86	MH	49.46	46.40
84.	Botad	96.58	MH	39.51	57.07
85.	Mendarda Mahal	96.65	MH	51.41	45.24
86.	Dharmpur	96.73	MH	49.89	46.84
87.	Dhandhuka	96.85	MH	49.67	47.18
88.	Balasinor	97.66	MH	60.87	36.79
89.	Bhavnagar	98.79	MH	43.80	54.99
90.	Manavdar	98.98	MH	47.72	51.26
91.	Okhamandal	99.82	MH	59.57	40.25
92.	Lunawada	99.97	MH	49.46	50.51
93.	Jamkandorna	100.64	ML	47.21	53.43
94.	Palitana	100.74	ML	46.03	54.11
95.	Lilia Mahal	100.95	ML	54.28	46.67
96.	Dohad	101.30	ML	53.10	48.20
97.	Sidhpur	101.79	ML	42.72	59.07
98.	Sagbara	101.94	ML	47.15	54.79
99.	Kotda Sangani	102.00	ML	48.78	53.22
100.	Mandvi	102.24	ML	55.16	47.08
101.	Kapadvanj	102.26	ML	56.30	45.96
102.	Jamjodhpur	103.02	ML	43.10	59.92
103.	Radhanpur	103.30	ML	60.98	42.32
104.	Kalol	103.32	ML	50.16	53.16
105.	Anjar	103.80	ML	46.90	56.90
106.	Nasvadi	103.97	ML	57.47	46.50
107.	Mahuva	104.02	ML	48.21	55.81
108.	Porbandar	104.08	ML	49.46	54.62
109.	Halvad	104.57	ML	55.49	49.08
110.	Limbdi	104.76	ML	56.67	45.14
111.	Khambhat	104.77	ML	63.64	41.13
112.	Vijapur	104.94	ML	49.51	55.43
113.	Chota Udaipur	105.21	ML	57.83	47.38
114.	Khedbrahma	105.25	ML	65.00	40.25
115.	Palanpur	105.63	ML	58.48	47.15
116.	Rajula	105.75	ML	54.02	51.73
117.	Devgadbaria	105.85	ML	43.15	62.70
118.	Jetpur	106.64	ML	38.64	68.00
119.	Gariadhar	106.72	ML	45.92	60.80
120.	Jhaloda	107.20	ML	47.76	59.44
121.	Keshod	107.22	ML	52.50	54.72
122.	Kundla	107.23	ML	47.28	59.95

(*Appendix V Contd.*)

	1	*2*		*3*	*4*
123.	Harij	107.41	ML	61.52	45.89
124.	Daskroi	107.47	ML	37.77	69.70
125.	Sayla	107.77	ML	59.13	48.64
126.	Gadhada	107.90	ML	53.04	54.86
127.	Lakhtar	108.29	ML	64.78	43.51
128.	Dediapada	108.73	ML	49.05	59.68
129.	Morvi	109.22	ML	53.48	55.74
130.	Jetpur Pavi	109.29	ML	49.13	60.16
131.	Paddhari	109.54	ML	53.70	55.84
132.	Vadgam	109.73	ML	58.02	51.71
133.	Jambusar	109.97	ML	51.41	58.56
134.	Bhiloda	109.99	ML	74.46	35.53
135.	Patan	110.24	ML	47.88	62.36
136.	Kutiyana	110.36	ML	57.07	53.29
137.	Nakhatrana	111.31	ML	61.55	49.76
138.	Meghraj	111.56	ML	75.22	36.34
139.	Maliya	111.72	ML	72.08	39.64
140.	Limkheda	112.07	ML	62.30	49.77
141.	Kunkavav Vadia	112.30	ML	50.11	62.19
142.	Dehgam	112.55	ML	55.65	56.90
143.	Chotila	112.86	ML	56.58	56.28
144.	Vanthali	112.86	ML	53.59	59.27
145.	Kadi	113.11	ML	50.98	62.13
146.	Kheralu	113.51	ML	49.89	63.62
147.	Ranavav	114.32	ML	50.11	64.21
148.	Muli	114.34	ML	58.83	55.51
149.	Shehera	114.40	ML	54.62	59.78
150.	Lalpur	115.13	ML	57.66	57.47
151.	Jafrabad	115.46	ML	59.24	56.22
152.	Bhesan Mahal	115.62	ML	61.07	54.55
153.	Vijaynagar Mahal	116.19	ML	86.90	29.23
154.	Bhanvad	116.46	ML	54.67	61.79
155.	Sanand	116.57	ML	58.21	58.36
156.	Jodiya	116.98	ML	62.12	54.86
157.	Dhari	117.26	ML	49.73	67.53
158.	Viramgam	117.29	ML	56.90	60.39
159.	Danta	119.16	ML	78.67	40.49
160.	Talaja	119.97	ML	51.36	68.61
161.	Kankrej	120.51	L	66.51	54.00
162.	Kalavad	120.57	L	53.86	66.71
163.	Bhachav	120.72	L	66.85	53.87
164.	Patan Veraval	121.33	L	47.55	73.78
165.	Malia	123.00	L	63.32	59.68
166.	Babra	123.01	L	59.73	63.28
167.	Santrampur	124.20	L	70.16	54.04
168.	Deodar	125.34	L	71.81	53.53
169.	Kodinar	125.97	L	60.38	65.59
170.	Santalpur	126.85	L	78.80	48.05

(*Appendix V Contd.*)

	1	2		3	4
171.	Jasdan	128.18	L	57.39	70.79
172.	Chansma	128.85	L	62.61	66.24
173.	Dasada	130.14	L	63.80	66.34
174.	Dhanera	130.48	L	69.84	60.64
175.	Talala	130.98	L	67.77	63.21
176.	Mangrol	134.45	L	57.34	77.11
177.	Visavdar	136.68	L	59.78	76.90
178.	Rapar	139.00	L	74.46	64.54
179.	Tharad	139.64	L	78.26	61.38
180.	Vav	139.95	L	77.99	61.96
181.	Deesa	140.19	VL	70.76	69.43
182.	Una	141.31	VL	66.85	74.46
183.	Kalyanpur	146.64	VL	75.61	71.03
184.	Sami	159.27	VL	86.75	72.52
	Mean	98.04		48.48	48.86

VH = Very High.
H = High.
MH = Moderately High.
ML = Moderately Low.
L = Low.
VL = Very Low.

BIBLIOGRAPHY

Bradburn, N. (1969). *The Structure of Psychological Well-being*, Aldine Press, Chicago.

Bulsara, J.F. (1970). *Patterns of Social Life in Metropolitan Areas*, Research Programmes Committee of the Planning Commission, Government of India.

Coastes, B.F., Johnston, R.J., and P.L. Knox, (1977). *Geography and Inequality*, Oxford University Press, Oxford.

Critchfield, H.J. (1975). *General Climatology*, Chapter XV on Human Comfort, Prentice Hall of India Pvt. Ltd., New Delhi.

Dave, Manjula, B. (1988). City Size and the Levels of Development in Gujarat, Ph.D. Thesis of the Gujarat University, Ahmedabad.

Desai, A.T. and K.M. Kulkarni, (1988). Four Tribal Villages of North Gujarat. A Study in Man-Environment Interaction (forthcoming issue of the Annals, *NAGI*).

Deshpande, C.D. (1948). *Western India – A Regional Geography*, Students Own Book Depot, Dharwar.

Dholakia, R.H. and V.K. Shah (1979). Changing Ranks of Talukas over a short period of Time – A Study of Gujarat, *Artha-Vikas*, Vol. XV, No. 1 (Jan-June), pp. 74-88.

Dikshit, K.R. (1974). *Geography of Gujarat*, National Book Trust, New Delhi.

Drewnowski, J. (1974). *On Measuring and Planning the Quality of Life*, Mouton, The Hague.

Dube, R.S. (1982). Social Well-being in Madhya Pradesh, *Trans. Inst. of Indian Geographers*, Vol. 4, No. 2, (July), pp. 169-174.

Gosal, G.S. and Gopal Krishna (1984). *Regional Disparities in Levels of Socio-economic Development in Punjab*, Vishal Publications, Kurukshetra.

Herbert, D.T. and D.M. Smith (1979). *Social Problems and The City: Geographical Perspectives*, Oxford University Press, New York.

Ilbery, B.W. (1984). Core-Periphery Contrasts in European Social Well-being, *Geography*, 69, pp. 289-302.

Jones Emrys and Eyles John (1977). *An Introduction to Social Geography*, Oxford University Press, London.

Knox, P.L. (1975). *Social Well-being : A Spatial Perspective*, Oxford University Press, London.

Kulkarni, K.M. (1984). Levels of Crowding and Social Well-being in Intra-urban Environments, *Annals of the NAGI*, Vol. IV, No. 2 (December), pp. 41-48.

Kulkarni, K.M. (1987). Patterns of Social Well-being in Micro-Regions of Gujarat – A Regional Analysis, *The Indian Journal of Social Work*, Vol. XLVII, No. 4 (January), pp. 389-398.

Kundu, A. (1980). *Measurement of Urban Processes – A Study in Regionalisation*, Popular Prakashan, Bombay.

McCracken, K.W.J. (1983). Dimensions of Social Well-being-Implications of Alternative Spatial Frames, *Environment and Planning*, A 1983, Vol. 15, pp. 579-592.

Pichholiya, K.R. (1983). Urbanisation in Gujarat, *Nagarlok*, Vol. XV. No. 3, (Sept), p. 51.

Smith, D.M. (1973). *The Geography of Social Well-being in the USA*, McGraw Hill, New York.

Smith, D.M. (1977). *Human Geography: A Welfare Approach*, Edward Arnold, London.

Smith, D.M. (1979). *Where the Grass is Greener : Geographical Perspectives on Inequality*, Croom Helm, London.

Terleckyj, N.E. (1970). Measuring Progress towards social Goals: Some Possibilities at National and Local Levels, *Management Science*, 16, B, pp. 765-78.

United Nations Research Institute for Social Development (1970). *Studies in the Measurement of Levels of Living and Welfare*, Report No. UNRISD/70/C.26 U.N.R.I.S.D., Geneva.

INDEX